Springer Series in Chemical Physics

Volume 112

The purpose of this series is to provide comprehensive up-to-date monographs in both well established disciplines and emerging research areas within the broad fields of chemical physics and physical chemistry. The books deal with both fundamental science and applications, and may have either a theoretical or an experimental emphasis. They are aimed primarily at researchers and graduate students in chemical physics and related fields.

More information about this series at http://www.springer.com/series/11752

Kaoru Yamanouchi · Luis Roso · Ruxin Li
Deepak Mathur · Didier Normand
Editors

Progress in Ultrafast Intense Laser Science

Volume XII

Editors
Kaoru Yamanouchi
Department of Chemistry
The University of Tokyo
Tokyo
Japan

Luis Roso
Pulsed Lasers Center
Villamayor, Salamanca
Spain

Ruxin Li
Shanghai Institute of Optics
and Fine Mechanics
Chinese Academy of Sciences
Shanghai
China

Deepak Mathur
Tata Institute of Fundamental Research
Mumbai
India

Didier Normand
CEA Saclay—IRAMIS
Gif-sur-Yvette
France

ISSN 0172-6218
Springer Series in Chemical Physics
ISBN 978-3-319-23656-8 ISBN 978-3-319-23657-5 (eBook)
DOI 10.1007/978-3-319-23657-5

Library of Congress Control Number: 2015950879

Springer Cham Heidelberg New York Dordrecht London

Printed on acid-free paper

Springer International Publishing AG Switzerland is part of Springer Science+Business Media (www.springer.com)

Preface

We are pleased to present the twelfth volume of Progress in Ultrafast Intense Laser Science. As the frontiers of ultrafast intense laser science expand rapidly, there continues to be a growing demand for an introduction to this interdisciplinary research field that is at once widely accessible and capable of delivering cutting-edge developments. Our series aims to respond to this call by providing a compilation of concise review-style articles written by researchers at the forefront of this research field, so that researchers with different backgrounds as well as graduate students can easily grasp the essential aspects.

As in previous volumes of PUILS, each chapter of this book begins with an introductory part, in which a clear and concise overview of the topic and its significance is given, and moves on to a description of the authors' most recent research results. All chapters are peer-reviewed. The articles of this twelfth volume cover a diverse range of the interdisciplinary research field, and the topics may be grouped into four categories: atoms, molecules, and clusters interacting in intense laser field (Chaps. 1–4), laser-induced filamentation and laser propagation (Chaps. 5 and 6), laser–plasma interaction and application (Chaps. 7 and 8), and ultrafast photo-induced processes of organic materials (Chap. 9).

From the third volume, the PUILS series has been edited in liaison with the activities of the Center for Ultrafast Intense Laser Science at the University of Tokyo, which has also been responsible for sponsoring the series and making the regular publication of its volumes possible. From the fifth volume, the Consortium on Education and Research on Advanced Laser Science, the University of Tokyo, has joined this publication activity as one of the sponsoring programs. The series, designed to stimulate interdisciplinary discussion at the forefront of ultrafast intense laser science, has also collaborated since its inception with the annual symposium series of ISUILS (http://www.isuils.jp/), sponsored by JILS (Japan Intense Light Field Science Society).

We would like to take this opportunity to thank all of the authors who have kindly contributed to the PUILS series by describing their most recent work at the frontiers of ultrafast intense laser science. We also thank the reviewers who have read the submitted manuscripts carefully. One of the co-editors (KY) thanks Ms. Chie Sakuta and Ms. Mihoshi Abe for their help with the editing processes. Last but not least, our gratitude goes to Dr. Claus Ascheron, Physics Editor of Springer-Verlag at Heidelberg, for his kind support.

We hope this volume will convey the excitement of ultrafast intense laser science to the readers and stimulate interdisciplinary interactions among researchers, thus paving the way to explorations of new frontiers.

Tokyo, Japan — Kaoru Yamanouchi
Salamanca, Spain — Luis Roso
Shanghai, China — Ruxin Li
Mumbai, India — Deepak Mathur
Saclay, France — Didier Normand

Contents

Contributors

Dimitri Batani CEA, CNRS, CELIA (Centre Laser Intense at Applications), University Bordeaux, Talence, France

Yohann Brelet Laboratoire d'Optique Appliquée, ENSTA-Paristech/CNRS/Ecole Polytechnique, Palaiseau, France

Giancarlo Bussolino Intense Laser Irradiation Laboratory, Istituto Nazionale di Ottica, Pisa, Italy

G.-Y. Chen Joint Quantum Institute, University of Maryland, College Park, USA

Arnaud Couairon Centre de Physique Théorique, CNRS, Ecole Polytechnique, Palaiseau, France

Pengji Ding Laboratoire d'Optique Appliquée, ENSTA-Paristech/CNRS/Ecole Polytechnique, Palaiseau, France

Aditya K. Dharmadhikari Tata Institute of Fundamental Research, Mumbai, India

Jayashree A. Dharmadhikari Department of Atomic and Molecular Physics, Manipal University, Manipal, India

D.B. Foote Joint Quantum Institute and Institute for Physical Science and Technology, University of Maryland, College Park, USA

Lorenzo Fulgentini Intense Laser Irradiation Laboratory, Istituto Nazionale di Ottica, Pisa, Italy

Antonio Giulietti Intense Laser Irradiation Laboratory, Istituto Nazionale di Ottica, Pisa, Italy

Leonida A. Gizzi Intense Laser Irradiation Laboratory, Istituto Nazionale di Ottica, Pisa, Italy; INFN, Sezione di Pisa, Italy

N.K. Gupta Theoretical Physics Division, Bhabha Atomic Research Centre, Mumbai, India

Hirokazu Hasegawa The University of Tokyo, Tokyo, Japan

W.T. Hill III Department of Physics, Joint Quantum Institute and Institute for Physical Science and Technology, University of Maryland, College Park, USA

Aurélien Houard Laboratoire d'Optique Appliquée, ENSTA-Paristech/CNRS/Ecole Polytechnique, Palaiseau, France

H. Jang Department of Physics, University of Maryland, College Park, USA

Tsuyoshi Kato Department of Chemistry, School of Science, The University of Tokyo, Bunkyo-ku, Tokyo, Japan

Petra Koester Intense Laser Irradiation Laboratory, Istituto Nazionale di Ottica, Pisa, Italy

Luca Labate Intense Laser Irradiation Laboratory, Istituto Nazionale di Ottica, Pisa, Italy; INFN, Sezione di Pisa, Italy

J. Lee Department of Physics, University of Maryland, College Park, USA

Yi Liu Laboratoire d'Optique Appliquée, ENSTA-Paristech/CNRS/Ecole Polytechnique, Palaiseau, France

Erik Lötstedt Department of Chemistry, School of Science, The University of Tokyo, Bunkyo-ku, Tokyo, Japan

Deepak Mathur Tata Institute of Fundamental Research, Mumbai, India

Katsumi Midorikawa RIKEN Center for Advanced Photonics, Wako, Saitama, Japan

Gaurav Mishra Theoretical Physics Division, Bhabha Atomic Research Centre, Mumbai, India

Sergey Mitryukovskiy Laboratoire d'Optique Appliquée, ENSTA-Paristech/CNRS/Ecole Polytechnique, Palaiseau, France

André Mysyrowicz Laboratoire d'Optique Appliquée, ENSTA-Paristech/CNRS/Ecole Polytechnique, Palaiseau, France

Hiroaki Nishimura ILE, University of Osaka, Osaka, Japan

Yasuhiro Ohshima Tokyo Institute of Technology, Tokyo, Japan

Ken Onda PRESTO, Japan Science and Technology Agency and Graduate School of Science and Engineering, Tokyo Institute of Technology, Yokohama, Japan

Guillaume Point Laboratoire d'Optique Appliquée, ENSTA-Paristech/CNRS/Ecole Polytechnique, Palaiseau, France

Kaoru Yamanouchi Department of Chemistry, School of Science, The University of Tokyo, Bunkyo-ku, Tokyo, Japan

Chapter 1
Image-Based Closed-Loop Control of Molecular Dynamics: Controlling Strong-Field Dissociative-Ionization Pathways

G.-Y. Chen, J. Lee, H. Jang, D.B. Foote and W.T. Hill III

Abstract In this chapter we will describe a genetic-algorithm search method employing velocity map imaging, 2D fitness functions and multiple search criteria that allow different parts of the control landscape to be examined. We will show how this approach can be exploited for both probing and controlling a rich array of dynamics in polyatomic molecules in a clear and transparent way. As an example of its power and versatility, we will control the bending vibration of CO_2 during dissociative ionization while searching two distinct parts of the control landscape. In so doing, we have tested the theoretical prediction that all solutions are optimal when the constraints are not too severe. Finally, we will show that multiple solutions render the dynamics more transparent, enabling important steps toward deciphering the solution to be made.

G.-Y. Chen
Department of Physics, University of Maryland, College Park, USA
e-mail: gychen@umd.edu

J. Lee · H. Jang
Joint Quantum Institute, University of Maryland, College Park, USA
e-mail: jlee.optics@gmail.com

H. Jang
e-mail: hjang@umd.edu

D.B. Foote
Joint Quantum Institute and Institute for Physical Science and Technology, University of Maryland, College Park, USA
e-mail: dbfoote@umd.edu

W.T. Hill III(✉)
Department of Physics, Joint Quantum Institute and Institute for Physical Science and Technology, University of Maryland, College Park, USA
e-mail: wth@umd.edu

K. Yamanouchi et al. (eds.), *Progress in Ultrafast Intense Laser Science XII*, Springer Series in Chemical Physics 112, DOI 10.1007/978-3-319-23657-5_1

1.1 Introduction

Phase and spectrally-shaped laser pulses obtained via a combination of closed-loop feedback and learning algorithms (e.g., genetic or evolutionary algorithms, collectively referred to as GA in this chapter) can be used to produce a so-called optimal control pulse (OCP). These OCPs can be remarkably, if not surprisingly, effective at controlling a quantum system, where it is difficult, if not impossible, to articulate the problem analytically a priori. Given a *suitable* fitness function, however, the system essentially learns what pulse is required to achieve the desired result. In contrast to traditional searches that tend to follow local gradients, GA searches can avoid getting trapped at local maxima. Theoretically, all solutions are optimal solutions if the constraints on the search space are not too severe; there are no suboptimal solutions [1]. Experimentally, obtaining the optimal solution depends strongly on the fitness function. To maximize product yield or any well-defined endpoint that can be measured directly, such as isolating a single harmonic during high-harmonic generation, a scalar usually servers quite well as the fitness function. There are cases of interest, however, where a scalar may not be most appropriate. In experiments employing velocity map images, for example, the most appropriate fitness function will be associated with regions of the image that could be both noncontiguous and two-dimensional (2D). In this chapter we will describe a GA-search method employing velocity map imaging and 2D fitness functions, and use them to study the "no suboptimal solutions" prediction. We will also show that this approach enables a rich array of dynamics to be studied and controlled, some of which are more transparent and less ambiguous than possible by other means.

Our target system, CO_2, has been the subject of strong-field dissociative ionization since the early 1990s, yet it continues to attract attention making it an ideal candidate for adaptive control studies. When exposed to intense near infrared radiation, CO_2 undergoes structural changes, which are partially responsible for the molecule acquiring a relatively large bending amplitude during strong-field dissociative ionization into singly and doubly charged atomic ions [2–6]. The bond length for enhanced ionization has been measured to be about 0.215 nm (4.06 a.u.) [5, 7] while the angular distribution for dissociative ionization is observed to be strongly peaked about the polarization axis obeying $\sim|\cos^{39}\theta|$ [8] for both linear and bent explosions, which corresponds to an alignment parameter of $\langle\cos^2\theta\rangle \simeq 0.95$. More recent experiments have focused on controlling the strong-field dissociative ionization. For example, Bocharova et al. [7] adjusted the width of an isolated pulse to control ionization and the bond length onset for explosion while Lozovoy et al. [9] and Chen et al. [10] employed adaptive control to find an OCP to enhance the bending vibration. The latter, to our knowledge, was the first image-based closed-loop molecular control demonstration and is the focus in this chapter. A similar approach has recently been used to study larger systems [11].

It is well known that OCPs are generally not unique; typically, more than one solution can be found with nearly the same effectiveness that might be explained by the theoretical prediction mentioned above that the search space can permit multiple solutions in different regions of the search space with none being suboptimal. While experimental results are consistent with this prediction, to our knowledge this has never been explored systematically experimentally. Such an exploration requires that the same dynamics be controlled via different GA approaches. In [10] we found the OCP with a search-space constraint on what possible OCPs could be created. To test the theoretical prediction, we will present results where we have searched for the OCP using two very different constraints involving different regions of the landscape.

This paper is organized as follows. The experimental setup and approach, including the 2D fitness function, will be described in Sect. 1.2. The experiments performed and their results will be outlined in Sect. 1.3 while their interpretation will be discussed in Sect. 1.4. Finally, we will add a few concluding remarks in Sect. 1.5.

1.2 Experimental Setup and Tools

Figure 1.1 sketches the experimental arrangement used for pulse shaping, strong-field dissociative ionization, GA-mediated landscape searching, and image capture. The experiments were driven by nearly transform limited, linearly polarized pulses produce by a Ti:sapphire laser system (not shown), which was adjusted to yield a kilohertz train of 1 mJ pulses with ~15 nm bandwidth. The pulses were shaped with a 128-elements liquid crystal spatial light modulator (CRI SLM-128) capable of phase and amplitude modulation. The spectral bandwidth of the pulse shaper was ~25 nm, when centered at 800 nm. After shaping, the beam was divided with a 80/20 beam splitter; the bulk of the energy was used for the experiment while the pulse was characterized with the remainder. The intensity in each arm was controlled independently with an (identical) attenuator composed of a half-wave plate and a polarizer. Pulses sent to the chamber, housing our 4π-image spectrometer [12], were focused with a spherical mirror (f.l. ~75 mm) to about a 7.2 μm radius, which was sufficient to create dissociative-ionization of CO_2 leading to doubly-charged atomic ions.

Nominally, the pulses were 50 fs in duration (corresponding to a bandwidth of ~15 nm) and centered at ~800 nm. In the single-pulse control scheme that we employed, the optimal pulse both controlled and assessed the dynamics, the latter via direct Coulomb-explosion imaging [12]. Consequently, we only shaped the phase (and not the amplitude) of the pulse to ensure there was sufficient intensity after shaping to ionize the parent molecular system multiple times, typically requiring an intensity $>10^{14}$ W/cm^2. Our experiments were performed with an ambient CO_2 pressure of $\sim 5 \times 10^{-8}$ Torr at laser intensities in the 0.5 to 2.7×10^{15} W/cm^2 range, which was sufficient intensity to remove five or six electrons prior to an energetic dissociation into $O^{l+} + C^{m+} + O^{n+}$ where $l, m, n = 1, 2$ subject to $l + m + n = 5, 6$. We imaged exclusively doubly charged O and C ions by gating the MCP voltage. An MCP gate resolution of 100 ns was sufficient to separate the first three charge states and to exclude protons, OH and N ions. Figure 1.2a is a typical image obtained after about 1 million laser shots. These composite images can be obtained with either the analog or digital camera.

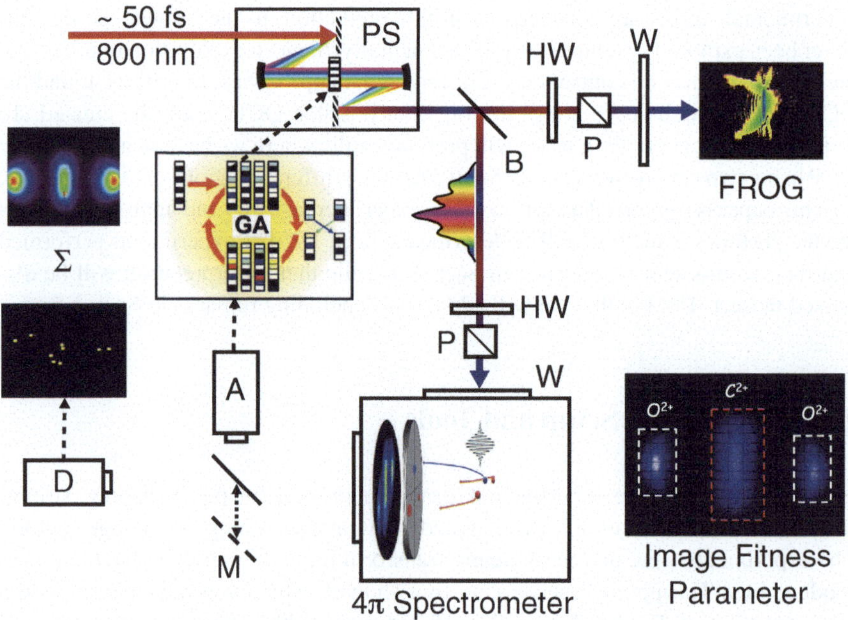

Fig. 1.1 Experimental arrangement for pulse shaping, image capture with our 4π-image spectrometer that uses a micro channel plate (MCP) to detect the ions [12], genetic algorithm feedback loop and pulse characterization using FROG [13]: *PS* spatial light modulator: based pulse shaper; *HW* half-wave plate; *P* polarizer; *W* window; *A* analog CCD camera used for the GA analysis; *D* high frame-rate digital CCD camera used for single-shot image capture and coincidence imaging; and *M* movable mirror to switch between cameras. Also shown are example images (clockwise from *top left*) a composite image, a FROG image of an OCP pulse, regions of the explosion image used as 2D fitness functions and a single frame image corresponding to a single laser shot generating four to six explosion events

1.2.1 Pulse Characterization

Pulses were characterized by transient-grating FROG (frequency-resolved optical gating [13]) running in polarization gating (PG) mode, where the nonlinear medium was a microscope cover slide of thickness ~0.2 mm. This characterization included retrieving the intensity and phase versus time and wavelength using standard FROG analysis software [16]. To ensure FROG characterization of the pulse was representative of the beam sent to the experimental chamber, we placed the same optical elements (viewport, beam attenuator, mirrors, etc.) in both arms. The only difference between the beams was that the FROG beam was transmitted through the 80/20 beam splitter while the experimental-chamber beam was reflected. Our tests showed that this did not cause any serious issues with determining the spectral phase.

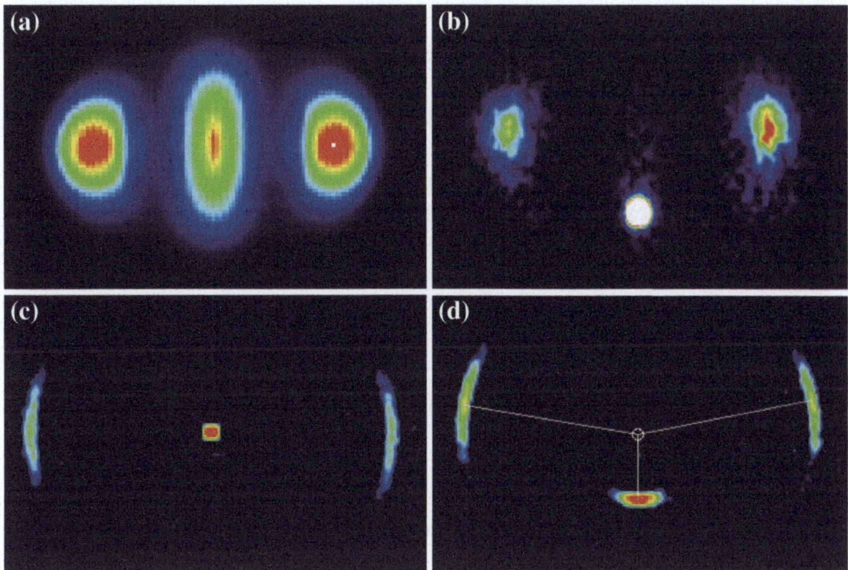

Fig. 1.2 Images of charge distributions of doubly charged ions, O^{2+} (outer lobes) and C^{2+} (central lobe), generated via explosions of CO_2^{n+} ($n = 5, 6$) obtained with the digital camera in Fig. 1.1 with the laser polarization axis horizontal on the page, which is parallel the plane of the MCP: **a** composite image charge distribution containing all doubly-charged trajectories (singly-charged ions not included) and channels (see text) after about 2.5×10^6 laser shots; **b** ions with their trajectories parallel to the MCP corresponding to a symmetric bent explosion of CO_2^{6+} with a specific bond angle extracted from (**a**) using the selective average techniques described in [14]; **c** triple-coincidence image of the angular distribution about the polarization axis in a plane parallel to the MCP for symmetric linear explosions CO_2^{6+} extracted from (**a**) using coincidence-imaging techniques (see text), which are described in [15] and references therein; **d** triple-coincidence image of the angular distribution about the polarization axis in a plane parallel to the MCP for symmetric bent explosions at a specific bond angle extracted from (**a**)

1.2.2 Image Spectrometer

The 4π-image spectrometer, which was inspired by the fast-ion imaging by Kanter et al. [17] and the photoelectron imaging detector of Helm et al. [18], has been described in detail elsewhere [12, 19]. Briefly, it consists of a uniform static electric extraction field (~250 V/cm) with an image-quality microchannel plate (MCP) to capture/image the ions created at the focal point. The focal point is located a perpendicular distance from the center of 40 mm dia. MCP circle. In operation, the polarization axis of the light is oriented perpendicularly with respect to the flight path to the MCP and parallel to the surface of the MCP. Atomic ions resulting from the explosions are swept towards the MCP by an extraction electric field. The spectrometer was run in two distinct modes: (i) summed acquisition, which we employed for GA closed-loop searches and (ii) single-shot acquisition, which we employed to analyze the dynamics. In the summed mode an analog CCD camera was used to capture images of the phosphor screen 640 × 480 pixel array. The camera frames were streamed to disk at 15 fps where they were summed in real time. The laser was ran at 1 kHz repetition rate so each camera frame contained a sum of ~67 laser shots, each of which inducing about 50 explosion events. Suitable images for GA analysis were typically captured within 1–2 min, corresponding to a composite image with a few tens of thousands of explosion events. In the single-shot mode, a fast-frame digital camera (730 fps) was used to capture images of the phosphor screen in 128 × 128 pixel array, each frame of which corresponding to a single laser shot. These camera frames were synchronized to the repetition rate of the laser, 700 Hz in this case, streamed to disk at 700 fps and stored individually. Figure 1.2a shows a composite image after $\sim 2.5 \times 10^6$ laser shots obtained either from the analog camera or summed images from the digital camera. When the digital camera is used it is possible to extract images in Fig. 1.2b–d as described below. In particular, Fig. 1.2b is associated with explosions that occur in a plane parallel to the detector face at a specific bond angle. Figure 1.2c, d show the angular distribution about the polarization axis for linear and bent explosions taken from [8]. In the bent case it is important to recognize that the bond angle is fixed.

The image in Fig. 1.2a is a 2D representation of a three-dimensional explosion; not all ion trajectories are parallel to the MCP. Consequently, the image is not a representation of the ions' momenta directly because of distortion by out-of-plane trajectories (see [19] for a more complete discussion). An extraction of the momentum distribution, which is necessary to determine the pre-explosion structure and any collateral dynamics (e.g., angular distributions), requires the various trajectories associated with different angles to be deconvolved via one of several possible techniques (see, for example [19–24]).

We used an alternative approach to extract this information from images taken with the digital camera. Each digital frame is associated with one-and-only-one laser shot, which enables correlation analysis—*image labeling* [14, 15], *triple coincidence* [8, 15] and *joint variance* [15]—to be used to select trajectories that were initially ejected in a plane parallel to the surface of the MCP. The images shown in Fig. 1.2 c and d, for example, were obtained from triple coincidence. The (x, y) positions of the ions relative to the center of the image are the true 2D momenta.

Genetic-Algorithm searches require images to be obtained as quickly as possible. One does not want to use any of the deconvolution or post-selection algorithms just mentioned in conjunction with a GA search because they are too slow to be incorporated into a feedback loop. It turns out that GA searches can be performed on raw images. This is possible if (1) specific areas on the image that are concomitant with the dynamics can be identified and (2) these areas can be used as the fitness function. Such areas become the fingerprints of the dynamics we wish to control, as detailed in Sect. 1.2.4. One example is the length of the C^{2+} central lobe in Fig. 1.2a. More acute bending necessarily involves a longer lobe so a fitness function could involve maximizing the lobe length. Once a solution is found, digital images are captured and triple-coincidence is used to quantify the bending.

We end this section by noting that the 4π-image spectrometer is similar to the family of image spectrometers based on recoil ion imaging, *COLTRIMS* (Cold Target Recoil Ion Molecular Spectroscopy) [25–27], which was developed independently at about the same time as ours is perhaps the most familiar.[1] The fundamental difference between the two approaches is in the way the electrons after the MCP are captured. We employ a phosphor screen and a CCD camera to determine the location of electrons ejected by the MCP rather than the position sensitive delay-line array used by *COLTRIMS*. This seemly small difference has significant consequences. Whereas as the spatial (i.e., energy) resolution in the plane of the MCP is nominally the same for the two approaches, *COLTRIMS* has higher energy (i.e., temporal) resolution in the time-of-flight (TOF) direction, which obviates the need for a deconvolution to determine momentum distributions. Higher resolution in this third dimension, however, comes at the expense of limiting the maximum number of simultaneous hits on the MCP per laser shot. For the standard delay-line detector with four axes (readouts), *COLTRIMS* is most effective for diatomic systems and about one molecular explosion per shot. The 4π spectrometer, by contrast, is able to handle many tens of simultaneous hits per laser shot and large molecules. Recently, a six-axis delay-line detector was introduced that increases the number of simultaneous hits significantly, enabling *COLTRIMS* to be employed with polyatomic systems as well [28]. The poorer temporal resolution in the TOF direction places limitations on the 4π spectrometer associated with overlapping trajectories (within tens of nanoseconds) and channel saturation (charges from certain channels appearing with each laser shot). The 4π spectrometer, however, has the advantage when it comes to GA searches and closed-loop feedback control. Many thousands of laser shots are required for GA searches because a discernible image of the explosion must be analyzed in realtime during the GA steps, described below, in order to reach convergence. Image capture with *COLTRIMS* is slower, by about an order of magnitude because generally only one molecular explosion per laser shot possible rather than many tens. This makes imaging with cameras more desirable (and cheaper) for GA searches. The 4π and *COLTRIMS* spectrometers are, thus, complementary.

1.2.3 GA Search Approach

Optimal fields obtained via GA searches depend critically on the *fitness function*, a measurable outcome strongly correlated with the desired dynamics (or final state) that the optimal field is to control. As already mentioned, in experimental searches, maximizing the fitness function must be done repeatedly, in real time, and as fast as possible so that the parameter space can be searched quickly. This is extremely

[1] While our current application does not include a cold molecular beam it could easily be added. The comparison we make between the two approaches is limited to the detection of the electrons ejected from the MCP, which is used to amplify the ion signal. Thus, we will refer to this family of spectrometers as *COLTRIMS* in this section.

important because depending on the number of *genes*—independent parameters—the search space can be extremely large. For example, there are about 2.5×10^{462} possible setting in our pulse shaper spanning 4π—128 elements each with 12-bit resolution; in this case there would be 128 genes. Things are not quite this bad because the SLM steps are nonlinear. However, even if we only scan over a 2π range, at 800 nm there are still about 1.5×10^{364} possible settings corresponding to about 700 steps between 0 and 2π. Depending on the dynamics, it is sometimes possible to reduce the space further by constraining the search as we discuss in Sect. 1.3. These constraints must be added carefully so as not to exclude the optimal solutions. Consequently knowledge of the dynamics and the systems is essential to run a GA successfully. As mentioned above, we use an area on the images as the fitness function, which falls into the class of 2D fitness functions. Two-dimensional fitness functions provide more detailed control of a specific channel while also allowing multiples channels to be controlled simultaneously. To control specific channels and dynamics, we must have an image *fingerprint* that identifies the desired process. We will discuss fingerprinting in the next section.

Our GA searches followed the traditional approach outlined in [29], with each of our generations consisting of 40 individuals. The population of each successive generation was obtained via fitness-proportional selection, recombination (one-point crossover), mutation, elitism and replacement. In fitness-proportional selection, the probability of an individual to have offsprings is proportional to its fitness. In one-point crossover, the genes of a chosen pair of individuals selected randomly are swapped. In mutation, every gene is given a chance, typically small but can be varied during the search, for its value to be altered by an amount chosen randomly. By piling up several mutations, a gene can migrate far away from initial value. The elitism operation passes a certain number of copies of fittest individual to the next generation to ensure that it will not be lost. Finally, individuals not satisfying a minimum intensity requirement to produce ionization are eliminated from the gene pool; the eliminated genes are replaced by others randomly generated and commensurate with the specific GA operation stage.

1.2.4 Channel Fingerprinting

Channel fingerprinting involves associating explosion channels with unique areas on the image. To that end, we simulated the three-body explosion by solving the coupled equations of motion,

$$m_i \ddot{r}_i = -\frac{e^2}{4\pi\varepsilon_0}\frac{\partial}{\partial r_i}\sum_{j\neq i}^{N}\frac{q_i q_j}{R_{ij}}, \tag{1.1}$$

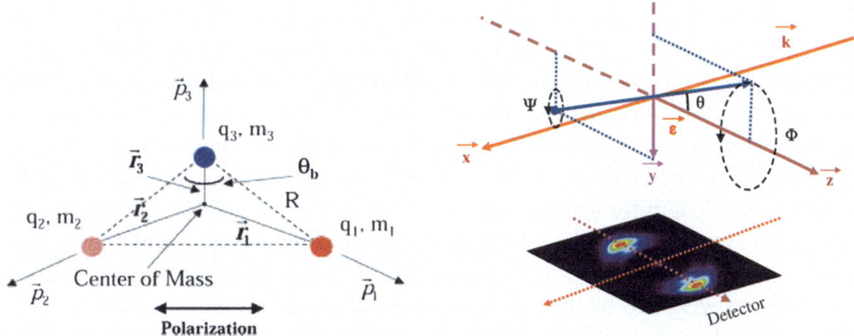

Fig. 1.3 Molecular geometry at the time of the explosion (*left*) and explosion geometry relative to the detector. *Left* q_i and m_i are the number of elemental charges, e ($q_i e$ is the charge of mass m_i), and the masses respectively, $\mathbf{r}_i$ and $\mathbf{p}_i$ are the distances from the center of mass and the initial momenta respectively, R are the charge separations (i.e., $R_{ij} = |\mathbf{r}_i - \mathbf{r}_j|$ and θ_b is the bond angle. *Right* $\boldsymbol{\varepsilon}$ and $\mathbf{k}$ are the polarization axis (parallel to $\mathbf{z}$) and the propagation vector (parallel to $\mathbf{x}$) respectively, θ is the angle between the dipole moment of the molecule and $\boldsymbol{\varepsilon}$, Φ the azimuthal angle the dipole moment of the molecule makes with $\boldsymbol{\varepsilon}$ and Ψ is the rotation of the molecule around its dipole moment (clearly only important when the molecule is bent

subject to the classical hamiltonian,

$$H = \sum_{i=1}^{N} \frac{1}{2} m_i \dot{r}_i + \frac{e^2}{4\pi\varepsilon_0} \sum_{i>j}^{N} \frac{q_i q_j}{R_{ij}}, \tag{1.2}$$

for a pure Coulomb explosion. In (1.1) and (1.2), R_{ij} are the critical radii (R_cs) for enhanced ionization [30, 31], ε_0 is the vacuum permittivity and the rest of the parameters are defined in Fig. 1.3. For the three-atom (CO_2) explosion of interest, $N = 3$. In principle, one could partially account for residual bonding in a mean field way by including a charge defect (σ), where $q \rightarrow q_{eff} = q - \sigma$ [5]. The Coulomb approximation near R_c is pretty good when five or six electrons are removed, however, so we did not use an effective charge in our simulations.

We modeled the dissociative ionization for the three channels that populate our images—(222), (221/122) and (212). The simulations were run in four steps, consistent with the recent experimental observation of Bocharova et al. [7]. Step one was the removal of three electrons from the parent CO_2. Step two was an expansion of the system to R_c.[2] Step three was the removal of two or three additional electrons (via enhanced ionization) creating CO_2^{5+} or CO_2^{6+}. Step four was the Coulomb explosion from R_c into the atomic ions. One generally observes a range of bond lengths centered about R_c in experiments, which could be due to a combination of issues such as molecules in the most intense part of the beam being ionized before R_c, motion

[2]Bocharova et al. [7] found that CO_2^{3+} is the intermediate state enabling the system to reach R_c and supporting the ensuing dynamics, not CO_2^{2+} suggested by theory [6, 32].

during ionization and potential curves not being purely Coulombic. Bocharova et al. [7] found CO_2 takes about 100 fs to reach R_c, which they report to be about 0.21 nm. Zhao et al. using 100 fs pulses found an R_c of about 0.24 nm. Thus, to accommodate these variations, we allowed the CO bond lengths to be distributed with a bi-Gaussian distribution about the measured R_c of [5]. Specifically, for lengths shorter than R_c the width of the Gaussian was chosen to be $0.2(R_c - R_e)$, where R_e is the equilibrium CO bond length for neutral ground state CO_2. For lengths longer than R_c, the width was set to $0.5(R_c - R_e)$. These choices were commensurate with the spreads observed by Chen et al. [10]. In the simulation, the two bond lengths were selected randomly and weighted by the half Gaussians independently, allowing for the possibility of one bond being longer than the other. We assumed that the probability for ionization was governed by the angle of the time-averaged dipole moment of CO_2. For linear CO_2 the angular distribution for linear and bent explosions was measured to be $\propto \cos^n \theta$, with n being $\sim$39 [8], where θ is the angle between the dipole moment of the molecule and the polarization axis of the laser. (The absolute value of a normalized distribution produces an alignment parameter, $\langle \cos^2 \theta \rangle$, for dissociative ionization of $\sim$0.95.)

A simulated image of the explosion was prepared by summing a large number of simulated explosion events with CO_2^{l+m+n+} randomly oriented (i.e., random values for θ, Ψ and Φ) consistent with the above constraints. Image generation had two components—Coulomb repulsion and field extraction. The atomic ions ($O^{l+} + C^{m+} + O^{n+}$) were allowed to expand under their mutual Coulomb force for about 5 ps at which time the Coulomb force was turned off because beyond this time it had negligible affect on the final momenta of the ions. At the same time, the ions were subjected to a static electric field pushing them toward the detector. Knowing the strength of the field and the distance to the detector, it was possible to determine where each ion will land on the MCP. This Monte Carlo simulation granted us the ability to relate the molecular dynamics to the contours of the image, thus identifying areas to be used as fitness functions for GA control—channel fingerprints, if you will. For symmetric bending of the (222) channel, the length of the central C^{2+} lobe corresponding to ions that have no momentum along the polarization axis is ideally suited as the fingerprint. This fingerprint shown in Fig. 1.4d; the associated fitness function for enhancing the bending amplitude corresponds to maximizing the length of this area. The response with and without running the GA search is shown in the two triple-coincidence images in Fig. 1.5. It is evident that both the C^{2+} lobe and the O^{2+} arc lengths are longer for the GA pulse. We point out that areas on the image where C^{2+} have momentum along the polarization axis identify an asymmetric explosion channel where either the CO bond lengths or the O ion charges are unequal. The hyperbolic area in Fig. 1.4a shows an appropriate fingerprint for such asymmetric explosions. In conjunction with the (222)-channel fingerprint, depending on the separation of the vertices, this hyperbolic fingerprint can be used to isolate the (221)/(122) channels from the (222) channel. Maximizing the relative strengths of one over the other allows the relative strength of these two

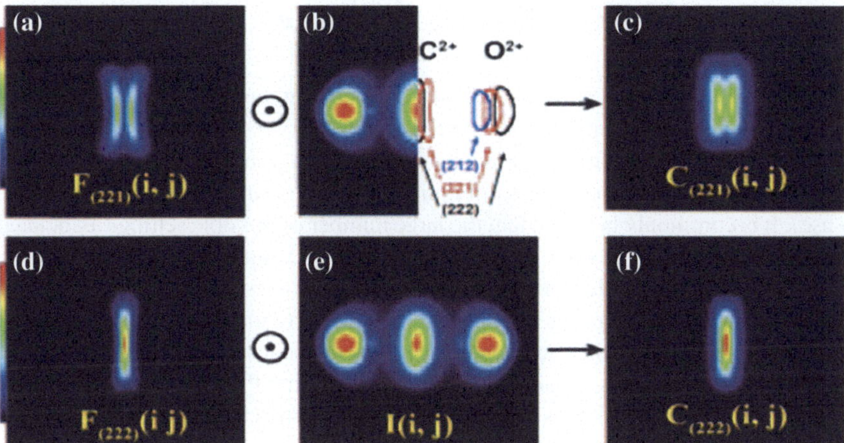

Fig. 1.4 Channel *fingerprints*, determined by simulations and represented as outlines in *panel b*, used as channel masks to enable area fitness functions for the (212), (221) and (222) channels. *Panel e* is a composite image of the doubly charged ions, $I(i, j)$, and *panel b* is an overlay of the fingerprints on $I(i, j)$. The isolated C^{2+} distributions associated with the channel masks shown respectively for simulated spectra ($F_{(221)}(i, j)$ and $F_{(222)}(i, j)$) and experimental spectra ($C_{(221)}(i, j)$ and $C_{(222)}(i, j)$). We do not isolate the (221) and (122) channels so $F_{(221)}(i, j)$ and $C_{(221)}(i, j)$ represent both channels

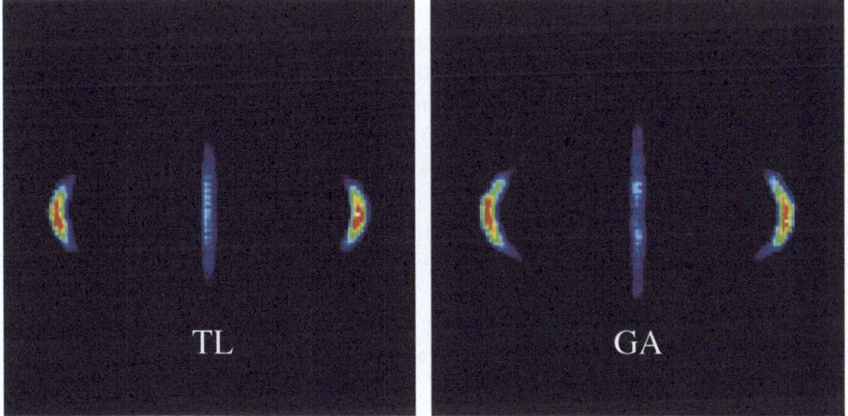

Fig. 1.5 Two triple-coincidence explosion images induced by TL (*left*) and GA-solution pulses. The TL image was obtained at an intensity of about 7×10^{14} W/cm^2 and the GA was obtained with the last peak in the train equivalent to about 6×10^{14} W/cm^2

channels to be controlled [33]. Using just one leg of the hyperbola enables isolation of the (221) from the (122) channel. In the remainder of this chapter we will focus on the (222) channel. The other channels will be treated in a separate publication.

1.3 GA Control of CO_2 Bending Amplitude

We will now present the results of three sets of experiments associated with enhancing the bending vibration amplitude of the (222) channel. Two sets of GA searches were run with two different fitness functions. We also ran control experiments with a single pulse that was either transform limited or chirped (in the positive or negative sense). It is commonly believed that the large number of possible settings associated with pulse shaper, $\sim 1.5 \times 10^{364}$ in our case, contains a high degree of redundancy. Consequently, to expedite finding optimal solutions, investigators have looked for ways to reduced the size of the search space. This is particularly true for complicated searches where convergence is an issue. This tends to be less of a concern for simple searches such as creating a transform limited pulse using the strength of the second harmonic signal as the fitness function. For our GA search for an OCP, we found that unless the space was reduced, the search did not converge. Reducing the space, however, must be done judiciously so as not to exclude the optimal solution from the space. Here, we present results where the space was reduced in two ways. The first, which was presented in [10], exploited the dynamics of the problem to reduce the number of genes from 128 to 4, leading to a search space of about 2.4×10^{11} possibilities. We will refer to this as the restricted search. In the second GA search we relaxed the restrictions of the first considerably and searched a space with about 4×10^{41} possible settings. We will refer to the second as the unrestricted search as it is thirty orders of magnitude larger than the first. The second experiment was done for two reasons: (1) to verify that the restricted OCP was part of the unrestricted space and (2) to test the no suboptimal solution hypothesis.

In [10] we showed that it is possible to choose a set of genes that reflects the nature of the dynamics we were trying to control. Specifically, we demanded that the solutions searched induce periodic kicks to the molecule. This was motivated by the idea that if we repeatedly hit the molecule at the right frequency with the right phase we could enhance the bending just like pushing a child on a swing. The basis set had just four genes that were related to terms in a 5th-order Taylor expansion of the spectral phase,

$$\varphi(\omega) = \sum_{n=0}^{5} \varphi_n(\omega_0) \frac{(\omega - \omega_0)^n}{n!}, \tag{1.3}$$

where $\varphi_n(\omega_0) \equiv \partial^n \varphi(\omega)/\partial \omega^n |_{\omega_0}$. While φ_0 and φ_1 determined the envelope phase and group delay, and thus are ignored in our search, the remaining four produced a train of pulses that are chirped and composed of peaks with increasing intensity. The left column of Fig. 1.6 shows two solutions that we will refer to as the *restricted* solutions. The traces are FROG reconstructions of the temporal intensities (black curves) and phases of the pulses.

In the second GA experiment we allowed the pixels to change independently. However, we did add two restrictions to eliminate some of the possible settings. First, we locked four adjacent pixels together, reducing the number of independent spectral units to 32; each pixel in a four-pixel unit was set to the same value during

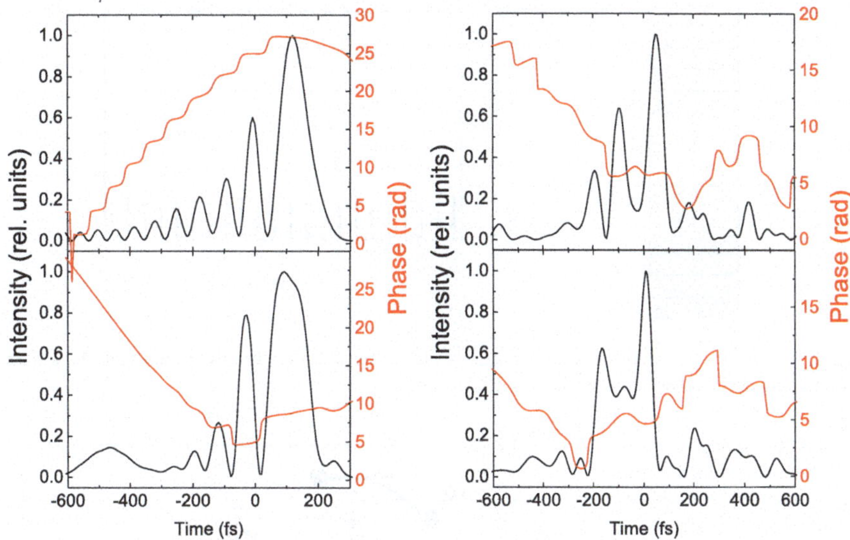

Fig. 1.6 Two *restricted*-GA solutions (FROG reconstructions) obtained with the Taylor-expansion gene set (*left*) and two *unrestricted*-GA solutions obtained with the 32-member gene set and a Gaussian cost functional. The *black curves* are the temporal intensities (*left axis*) and the *red curves* are the temporal phases (*right*)

the search. We also reduced the number of phase steps to 20; a four-pixel unit was allowed to make changes in increments of $\pm\pi/10$.[3] Even with only 20 steps, the phase can change drastically from generation to generation so we imposed an impedance to retard how quickly the phase for a particular unit can evolve. Specifically, we limited the range over which the phase of a given four-pixel unit could change in one generation to fall within $\pm\delta$ by adding a Gaussian (with a FWHM $= \delta$) cost functional centered at the current value of φ as shown graphically in Fig. 1.7; small changes are more probable than large changes. The GA solutions found were obtained with $\delta \sim \pi/4$. Two solutions are shown in the right column of Fig. 1.6.

As mentioned in Sect. 1.2.2 once an OCP is found, we exploit triple-coincidence images (similar to those of Fig. 1.5) to determine the bond length distribution. We extract the far-field angles associated with the bending of the molecule from these images. The molecular bond angle, θ_b, defined in the inset of Fig. 1.8, is determined numerically by solving the equations of motion (1.1) and (1.2) as described by Chen et al. [10]. The θ_b values are then plotted as shown in Fig. 1.8. We characterize the magnitude of the bending by the parameter $\Delta_{1/2}\theta_b$, the HWHM of the bending

[3]The number of steps was chosen to ensure the searches would converge in a reasonable amount of time. We did not explore the sensitivity of the OCP to the number of pixels in a unit nor the number of steps between 0 and 2π.

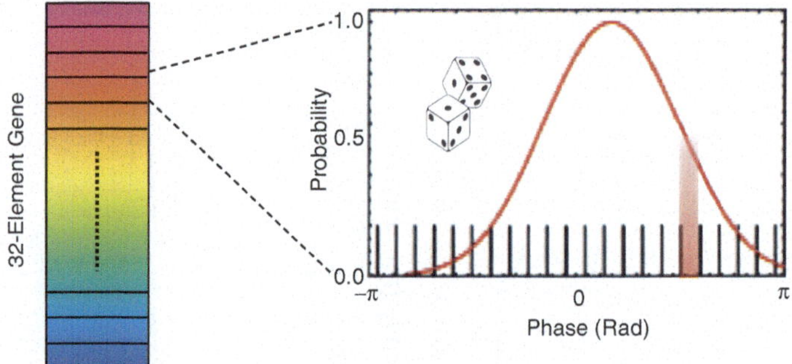

Fig. 1.7 The 20-step phase mutation range modified by a Gaussian cost functional. *Note* The Gaussian width (FWHM = δ, see text) is not to scale

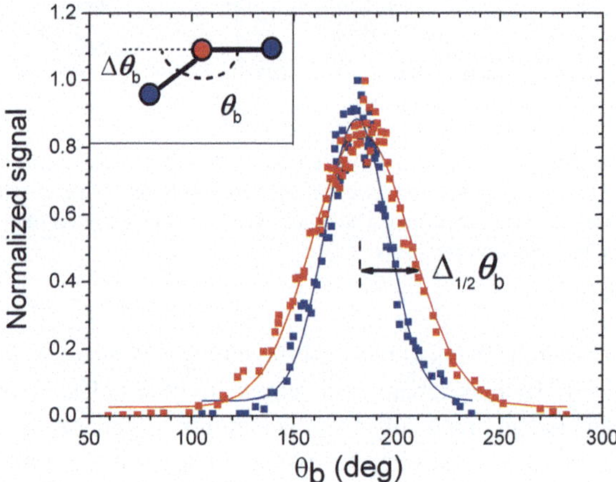

Fig. 1.8 Bond angle distributions obtained from triple-coincidence images as described in [10]. The *red dots* are the bond angle distribution of the optimal pulse taken with the OCP shown in the lower *right panel* of Fig. 1.6 and the *blue dots* are those of a 50 fs TL pulse. The *solid curves* are Gaussian fits to the data. The *inset* defines the angles

distribution in Fig. 1.8. We compare values of $\Delta_{1/2}\theta_b$ for the two GA and control experiments in Fig. 1.9. This figure is consistent with Fig. 3 of [10] where the GA value for $\Delta_{1/2}\theta_b$ is plotted at an effective intensity, which is determined by a TL pulse with a pulse width and pulse energy equivalent to that of the last peak in the GA solution. Both the width and effective energy of the GA peak are determine by FROG reconstruction. We then compare the bending due to the GA pulse to that of a TL that has a pulse energy equivalent to the total pulse energy of the entire GA pulse. For these experiments, the total pulse energy of the GA pulses was about 50 μJ. A 50 μJ TL pulse corresponds to a TL pulse with an intensity of about 9.4×10^{14} W/cm^2, which is the second TL pulse from the left in Fig. 1.9. The effective intensities for the GA pulses are given in Table 1.1. The two restricted solutions from [10] are also included in this plot.

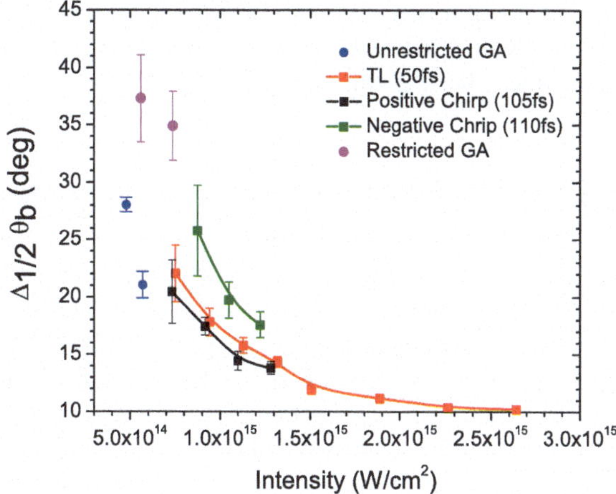

Fig. 1.9 Bond angle distributions for all four OCPs, 50 fs TL pulses as a function of intensity, and chirped pulses as a function of intensity

Table 1.1 Pulse parameters and bending results for the restricted ([10]) and unrestricted GA solutions

	A (upper)		A (lower)		B (upper)		B (lower)	
Peak #	1	2	1	2	1	2	1	2
τ_i	100	47	128	52	54	43	52	NA
I_i	7.5	4.2	5.6	4.3	5.7	3.4	4.8	3.2
Δt_{ij}	130		139		144		159	
$\Delta_{1/2}\theta_b$	35°		38°		21°		28°	
% Change	46 %		56 %		19 %		57 %	

A (B) corresponds to the restricted (unrestricted) solution in the left (right) column of Fig. 1.6; peak # 1 (2) corresponds to the last (second to last) peak in the solution pulse trains; τ_i (fs) are the temporal width of the peaks; Δt_{ij} (fs) are the time interval between the last two primary peaks; I_i are the intensity of the peaks in units of 10^{14} W/cm^2; $\Delta_{1/2}\theta_b$ is the bending parameter (see text) with an uncertainty of $\pm 1°$ to $2°$; and % change is the enhancement given by $100 \times (\Delta_{1/2}\theta_{b-GA} - \Delta_{1/2}\theta_{b-TL})/\Delta_{1/2}\theta_{b-TL}$ as described in the text

1.4 Discussion

Before we discuss the effectiveness of the GA solutions (OCPs) achieving their goals, we will first compare the structure of the GA solutions. We first note that all four solutions are composed of a set of well-defined peaks. For the restricted GA solution, the train of pulses was a result of an alternation in even and odd orders of a spectral phase [34]. The fact that we see a series of peaks for the unrestricted GA search, lends credence to our initial thought that the OCP should contain periodic kicks by a train of pulses. However, clearly a long train is not necessary because the solutions are not identical. Neither the pulse widths nor their separations are the same. The number of pulses varies drastically from solution to solution and even the phase structure is distinct for each pulse. What does seem to be critical is that the last peak in the chain contain considerably more energy resulting in higher intensity. We showed in [10] that reversing this trend, by putting the most intense peak first, greatly reduces the enhancement. Clearly, because we can reach a solution with as few as two peaks, a train, albeit short, is required.

1.4.1 GA-Solution for Bending Enhancement

We will now look at the degree to which the GA solutions enhanced the bending. We begin by summarizing the two restricted results. In [10] we determined the effectiveness by comparing $\Delta_{1/2}\theta_b$ produced by GA pulses to that produced by TL pulses with the same width as the last peak in the chain of the GA solution and the same total peak energy. For the restricted solutions, the widths of the last peak varied in duration from 100 to 128 fs (see Table 1.1). Transform-limited pulses of those widths, not shown in Fig. 1.9, produced $\Delta_{1/2}\theta_b \sim 24°$ while the best GA pulse produced $\Delta_{1/2}\theta_b \sim 37.5°$ for a maximum enhancement of ~56 %.[4] The other GA solution enhanced the bending by about 46 %. Compared with the TL pulse plotted at an intensity of 9.4×10^{14} W/cm^2, the two unrestricted GA solutions enjoyed enhancements of 19 and 57 %. We note that the uncertainties in $\Delta_{1/2}\theta_b$ are in the ±1° to 2° range.

It is evident that the pulse intensity, pulse width, and the spacing between the last two pulses tend to play a role in enhancing CO_2 bending during dissociative ionization for the symmetric (222) channel. As discussed in [10] bending is inversely proportional to pulse intensity (energy) as shown in Fig. 1.9. Long pulses were shown to lead to more enhancement than shorter. Both of these observations are likely caused by the molecule spending more time in the di- or tri-cation states, with contours that are sensitive to the intensity of the laser (see [6, 32]), before enhanced ionization occurs. The longer pulses give the system more time to respond to the changing shape of the potential surfaces [7]. Figure 1.9 also shows bending is sensitive to the pulse chirp. Negative chirp enhances bending for a pulse with the same width and intensity whereas a positive chirp seems to play a much less critical role. This is due to the curvature of potential surfaces enabling resonance to be maintained longer for negative chirps, which has been observed in other systems as well [35].

Focusing more directly on the GA (OCP) solutions (see Table 1.1 and Fig. 1.9), we notice some of the same trends. Bending is again inversely proportional to intensity. It scales with the duration of the final peak. This is obvious when comparing the restricted solution against the unrestricted solution. The width of the restricted, which is at least twice as large as the unrestricted, are much more effective at enhancing the bending. A similar case can be made between the two unrestricted solutions. While the widths are nominally the same, there is a side peak between the last two major peaks that effectively lengthens the last peak and could contribute to it being

---

[4] To generate the longer pulses we placed a positive chirped on the pulse. Figure 1.9 shows that a positive chirp does not enhance bending.

a bit more effective. The spacing between the two final peaks is another clue to the control. In our searches, we found more enhancement when the separation was close to an integer multiple of 50 fs, the bending vibrational period of the ground state of CO_2; the restricted solutions are both closer to 150 fs than the unrestricted solutions and the lower unrestricted solution is closer than the upper solution in Fig. 1.7.

It is also important to consider the chirp of the final peaks. In general, when the phase is flat across a peak it is nominally unchirped (transform limited). When the phase is parabolic across the peak with the vertex up (down) it will be positively (negatively) chirped. For the restricted-GA solutions, the upper solution in Fig. 1.7 is more positively chirped than the lower while the lower solution enhances the bending more. The same is true for the unrestricted-GA solution, but in this case the difference between the two is more stark. It is important to note that the difference between their relative effectiveness is also more pronounced than the restricted case. This is consistent with what we observed with a single, negatively-chirped pulse.

1.4.2 Suboptimal Solutions

As the last topic of this chapter we return to the no suboptimal solution prediction. Is it consistent with our experimental observations? Upon first glance it would appear no, because the four GA solutions (OCPs) did not achieve the same degree of bending as measured by $\Delta_{1/2}\theta_b$, which ranges from 21° to 28° for the unrestricted search and 35° to 37.5° for the restricted search. Taking into account the error bars, the restricted solutions are consistent with each other but the unrestricted are on the fringe of agreement at best. There is still the issue, however, the two searches do not achieve the same level of efficiency. Things do not look any better if the percentage enhancement is used as a measure—a range of 46 to 56 % for the to restricted solutions and 19 to 57 % for the unrestricted. The error bars are larger but we end up with the same conclusion. To drive this point home, we also found solutions, i.e., GA search converged maximizing the fitness function, but bending was not enhanced. These observations begs two questions: (1) why does it appear there are suboptimal solutions, (2) why does the restricted search do better than the unrestricted?

Taking the second question first, we remind the reader that the restricted search was designed specifically to give the molecule a period kick, which evidently is close to the right solution. While the fitness function asked the search to maximize its length, most (but not all) solutions also maximized the bending. The unrestricted was just that and had to search the landscape for regions to stretch. We suspect that stretching the C^{2+} lobe could be done in ways inaccessible to the restricted search. To backup this conjecture, we created triple-coincidence images selecting parts of the C^{2+} lobe where the C^{2+} had some momentum along the polarization axis using both the restricted and unrestricted solutions. As mentioned in Sect. 1.2.4, these images will pick out asymmetric explosions. While these images only contain doubly charged ions, the atomic ions can come from both CO_2^{6+} and CO_2^{5+} explosions will. Hence, the C^{2+} and O^{2+} resulting from a GA search can be due to (122), (212), (221) or (222) channels. We discovered that the image associated with the restricted-GA searches had considerably less contamination from CO_2^{5+} explosions than did the images associated with the unrestricted-GA searches. Whereas the C^{2+} arc length resulting from restricted-GA searches was due mostly to the (222) channel those from unrestricted-GA searches had a high degree of contamination from the asymmetric channels, especially (221) and (122). Even though the (222) channel populated both searches, the relationship between the length of the C^{2+} central lobe and the O^{2+} arc length was less isomorphic. That is, the C^{2+} could be extended because of the asymmetric channels without the (222) channel bending more. This is because of the extended focal volume allowing explosions occurring at the edge of the focal volume to produce ions that end up along the central C^{2+} lobe within the fingerprint. These ions will have some momentum along the polarization axis and thus represent asymmetric explosions. These ions come primarily from CO_2^{5+} explosions. The details will be discussed in more detail in a future publication. Suffice it to say, the net result is that the unrestricted searches can maximize the central lobe in ways that do not require the (222) channel to be involved giving solutions the appear to be suboptimal. This seems to occur much less frequently on the restricted landscape.

Returning to the first question, our evidence is consistent with the no suboptimal prediction. We asked the GA to maximize the length of the C^{2+} lobe. When the search converged on a solution it tends to be rather consistent. While this prediction needs to be studied more systematically, we do not disagree with it. What our observations do point out is that for complicated GA searches, more sophisticated fingerprints and fitness functions are required. The 2D approach suggested in this chapter are ideally suited for that purpose.

1.5 Conclusion

In this chapter we have presented a closed-loop GA search approach based on 2D fitness functions. This approach provides much more access to complicated dynamics associated with polyatomic systems than is possible with scalar fitness functions. We showed that careful fingerprinting of channels associated with the dynamics under study is required both to reveal the underlying physics and to control the dynamics upon which it is based. For bending during strong-field dissociative ionization of CO_2, we were able to link the effectiveness of its control to four distinct parameters that are summarized in Table 1.2. When the GA search converged, it maximized the fitness function, however, the fitness function was only a necessary condition for maximizing bending. As a result, other dynamics contributed to maximizing the fitness function making the suboptimal prediction appear to be violated. To circumvent

Table 1.2 Summary of CO_2 bending responses to pulse parameters during dissociative ionization

Parameters	Bending response
Intensity	Inversely proportional but nonlinear
Width	Directly proportional over the range tested, 50–150 fs
Separation	Larger response near an integer multiple of 50 fs
Chirp	Larger response for negative chirp

this problem, one needs to design more complete fitness function, which is possible with the addition of secondary and tertiary search conditions—areas on the image. To go to the next level, a fingerprint leading to a fitness function that is both necessary and sufficient must be incorporated in the GA search, where the fitness function and the dynamics under study are isomorphic.

Acknowledgments We thank Dr. G.M. Menkir for helpful discussions and technical support in the early stages of this work and Mr. B. Crist for technical support with the simulations. This work was supported by NSF Grant No. PHY0902221.

References

1. H.A. Rabitz, M.M. Hsieh, C.M. Rosenthal, Science **303**(5666), 1998 (2004)
2. C. Cornaggia, M. Schmidt, D. Normand, J. Phys. B: At. Mol. Opt. Phys. **27**, L123 (1994)
3. C. Cornaggia, F. Salin, C.L. Blanc, J. Phys. B: At. Mol. Opt. Phys. **29**, L749 (1996)
4. A. Hishikawa, A. Iwamae, K. Yamanouchi, J. Chem. Phys. **111**, 8871 (1999)
5. K. Zhao, G. Zhang, W.T. Hill III, Phys. Rev. A **68**, 063408 (2003)
6. Y. Sato, H. Kono, S. Koseki, Y. Fujimura, J. Am. Chem. Soc. **125**, 8019 (2003)
7. I. Bocharova, R. Karimi, E.F. Penka, J.P. Brichta, P. Lassonde, X. Fu, J.C. Kieffer, A.D. Bandrauk, I. Litvinyuk, J. Sanderson, F. Légaré, Phys. Rev. Lett. **107**, 063201 (2011)
8. K. Zhao, W.T. Hill III, Phys. Rev. A **71**, 013412 (2005)
9. V.V. Lozovoy, X. Zhu, T. Gunaratne, D. Harris, J. Shane, M. Dantus, J. Phys. Chem. A **112**, 3789 (2008)
10. G.Y. Chen, Z.W. Wang, W.T. Hill III, Phys. Rev. A **79**, R011401(R) (2009)
11. E. Wells, C.E. Rallis, M. Zohrabi, R. Siemering, B. Jochim, P.R. Andrews, U. Ablikim, B. Gaire, S. De, K.D. Carnes, B. Bergues, R. de Vivie-Riedle, M.F. Kling, I. Ben-Itzhak, Nat. Commun. **4**, 2895 (2013)
12. J. Zhu, W.T. Hill III, J. Opt. Soc. Am. B **14**, 2212 (1997)
13. R. Trebino, *Frequency-Resolved Optical Gating: The Measurement of Ultrashort Laser Pulses* (Kluwer Academic Publishers, Boston, 2000)
14. K. Zhao, G. Zhang, W.T. Hill III, Opt. Express **9**, 42 (2001)
15. W.T. Hill III, K. Zhao, L.N. Elberson, G.M. Menkir, in *Progress in Ultrafast Intense Laser Science I*, ed. by K. Yamanouchi, S.L. Chin, P. Agostini, G. Ferrante (Springer-Verlag, Berlin, 2006), Springer Series in Chemical Physics, pp. 59 – 75
16. (2014). URL http://www.swampoptics.com/
17. E.P. Kanter, P.J. Cooney, D.S. Gemmell, K.O. Groeneveld, W.J. Pietsch, A.J. Ratkowski, Z. Vager, B.J. Zabransky, Phys. Rev. A **20**(3), 834 (1979)
18. H. Helm, N. Bjerre, M.J. Dyer, D.L. Huestis, M. Saeed, Phys. Rev. Lett. **70**(21), 3221 (1993)

19. K. Zhao, T. Colvin, W.T. Hill III, G. Zhang, Rev. Sci. Instrum. **73**, 3044 (2002)
20. C. Bordas, F. Paulig, H. Helm, D. Huestis, Rev. Sci. Instrum. **67**, 2257 (1996)
21. G.A. Garcia, L. Nahon, I. Powis, Rev. Sci. Instrum. **75**, 4989 (2004)
22. G.M. Roberts, J.L. Nixon, J. Lecointre, E. Wrede, J.R.R. Verlet, Rev. Sci. Instrum. **80**, 053104 (2009)
23. E. Wells, J. McKenna, A.M. Sayler, B. Jochim, N. Gregerson, R. Averin, M. Zohrabi, K.D. Carnes, I. Ben-Itzhak, J. Phys. B: At. Mol. Opt. Phys. **43**, 015101 (2010)
24. T. Gerber, Y. Liu, G. Knopp, P. Hemberger, A. Bodi, P. Radi, Y. Sych, Rev. Sci. Instrum. **84**, 033101 (2013)
25. J. Ullrich, R. Moshammer, R. DÖrner, O. Jagutzki, V. Mergel, H. Schmidt-BÖcking, L. Spielberger. J. Phys. B: At. Mol. Opt. Phys. **30**, 2917 (1997)
26. R. DÖorner, V. Mergel, O. Jagutzki, L. Spielberger, J. Ullrich, R. Moshammer, H. Schmidt-BÖocking, Phys. Rep.**330**, 95 (2000)
27. J. Ullrich, R. Moshammer, A. Dorn, R. Dörner, L.P.H. Schmidt, H. Schmidt-öcking, Rep. Prog. Phys. **66**, 1463 (2003)
28. M. Pitzer, M. Kunitski, A.S. Johnson, T. Jahnke, H. Sann, F. Sturm, L.P.H. Schmidt, H. Schmidt-Böcking, R. Dörner, J. Stohner, J. Kiedrowski, M. Reggelin, S. Marquardt, A. Schießer, R. Berger, M.S. Schöffler, Science **341**, 1096 (2013)
29. K. Sastry, D. Goldberg, G. Kendall, in *Search Methodologies: Introductory Tutorials in Optimization and Decision Support Techniques*, ed. by E. Burke, G. Kendall, Business Media, LLC (Springer Science, 233 Spring Street, New York, NY 10013, 2005), Chap. 4:, pp. 97–125
30. S. Chelkowski, A.D. Bandrauk, J. Phys. B: At. Mol. Opts. Phys. **28** (1995)
31. H. Yu, A.D. Bandrauk, Phys. Rev. A **56**, 685 (1997)
32. H. Kono, Y. Sato, M. Kanno, K. Nakai, T. Kato, Bull. Chem. Soc. Jpn. **79**, 196 (2006)
33. G.Y. Chen, H. Jang, J. Lee, W.T. Hill III, To be published (2014)
34. K. O'Keeffe, T. Robinson, S.M. Hooker, J. Opt. **12**, 015201 (2010)
35. N. Schirmel, N. Reusch, P. Horsch, K.M. Weitzel, Faraday Discuss. **163**, 461 (2013)

Chapter 2
Classical Trajectory Methods for Simulation of Laser-Atom and Laser-Molecule Interaction

Erik Lötstedt, Tsuyoshi Kato, Katsumi Midorikawa and Kaoru Yamanouchi

Abstract The classical trajectory Monte Carlo method applied to the simulation of many-electron atoms and molecules in intense laser fields is reviewed. Two ways to solve the problem of constructing a stable, many-body ground state in classical mechanics are presented: (i) Fermionic molecular mechanics and (ii) interaction via soft-core Coulomb potentials. Five examples of the application of classical trajectory methods to the simulation of laser-driven atomic and molecular systems are introduced, ranging from the non-sequential double ionization of He and inner-shell electron ejection in C to the ionization and dissociation dynamics of H_3^+ and proton recollision in ^{15}NH.

2.1 Introduction

When a many-electron atom or molecule is exposed to an intense laser field (with a wavelength of typically 800 nm), in general several electrons respond simultaneously to the force of the laser field. Some electrons may individually absorb energy from the laser field, and later transfer part of the absorbed energy to other electrons through the Coulomb interaction. An extreme example of this kind of energy redistribution is the nonsequential double ionization (NSDI) process [1–7]: after field-induced ejection of one electron, the electron is accelerated by the laser field in such a way that it

E. Lötstedt (✉) · T. Kato · K. Yamanouchi
Department of Chemistry, School of Science, The University of Tokyo,
7-3-1 Hongo, Bunkyo-ku, Tokyo 113-0033, Japan
e-mail: lotstedt@chem.s.u-tokyo.ac.jp

T. Kato
e-mail: tkato@chem.s.u-tokyo.ac.jp

K. Yamanouchi
e-mail: kaoru@chem.s.u-tokyo.ac.jp

K. Midorikawa
RIKEN Center for Advanced Photonics,
2-1 Hirosawa, Wako, Saitama 351-0198, Japan
e-mail: kmidori@riken.jp

K. Yamanouchi et al. (eds.), *Progress in Ultrafast Intense Laser Science XII*,
Springer Series in Chemical Physics 112, DOI 10.1007/978-3-319-23657-5_2

returns close to the atomic or molecular core and collides inelastically with another, still bound, electron [8, 9]. If the energy transfer is large enough, the bound electron may be ejected from the atom or molecule, resulting in a final state with two free electrons. It is also possible for the returning electron to knock out not only one, but several bound electrons [10–19], which is termed nonsequential multiple ionization (NSMI). In recent years, it has become clear that the motion of several electrons in the laser field has to be accounted for in strong-field processes, as evidenced by the observation of multi-electron signatures in high-order harmonic generation [20–22].

In the case of molecules, not only the sharing of energy among electrons, but also the coupling between electrons and nuclei are interesting. As exemplified in recent investigations on $D_3{}^+$ [23–25], ethanol [26], formic acid [27] and small hydrocarbon molecules [28, 29], the outcome of a unimolecular dissociation reaction in an intense laser field is strongly dependent on how the electrons within a molecule are excited. The electrons in the molecule directly absorb energy from the laser field, and become excited and/or ejected from the molecule. As a result, the interatomic forces within the molecule change, leading to different kinds of dissociation and rearrangement reactions. Ultimately, we would like to realize the *control* of the dissociation dynamics of a laser-driven molecule by controlling the intramolecular electron dynamics. This kind of strong-field control has been successfully implemented mainly for simple diatomic molecules such as H_2 [30, 31] or DCl [32], but also for polyatomic molecules like C_2H_2 [33–35], C_2H_4 [36], and C_2H_5OH [26, 37–41].

For theory, it has not been an easy task to model laser-driven many-particle dynamics. As we have discussed above, a necessary requirement for any theory aiming to describe energy exchange among many particles is that more than one particle need to be allowed to move. Thus, we must go beyond the single active electron approximation. In quantum mechanics, this means that we should solve the time-dependent many-body Schrödinger equation (TDSE). Since the TDSE is a partial differential equation depending on $3N$ spatial variables where N is the number of particles, numerical solutions of the TDSE are extremely demanding. To date, exact numerical solutions of the TDSE have been obtained only for laser-driven He ($N = 2$) [42, 43]. To go beyond two-electron systems, approximate methods such as time-dependent density functional theory (TDDFT), [44–47], time-dependent multi-configuration expansions of a wave function [48–57], or R-matrix theory [58–60] are required.

In view of the excessive computational requirements to solve the TDSE for more than two laser-driven particles, it is natural to look for alternative ways of simulating laser-driven atoms and molecules. One such alternative method is to use classical mechanics for the description of the time-dependent motion of the particles. The motivation for using classical mechanics to simulate an inherently quantum mechanical system such as an atom or a molecule is essentially practical: While the TDSE, being a *partial* differential equation of $3N + 1$ variables, cannot be easily solved numerically, Newton's equation is an *ordinary* differential equation which can be straightforwardly solved even for a large number of interacting particles. The idea is to apply concepts from classical plasma and cluster physics [61] in order to understand the behavior of many-electron atoms and many-electron molecules exposed to an intense laser field. An atom or molecule can be viewed as a "sub-nano plasma" consisting of a group of free electrons and nuclei interacting via two-body potentials.

A further benefit from using classical mechanics for the simulation is that it is easy to understand the physical meaning of the calculated results. Since a classical trajectory contains information of both the positions and momenta of all particles at any time, it is possible to follow the dynamics during the interaction with the laser pulse, all the way from the unperturbed initial state to the final state containing free electrons or dissociated fragments (in the case of a molecule). Of course, one trajectory alone does not provide any useful information, and therefore, one has to calculate a large number of trajectories, with initial values sampled from a distribution of initial state configurations. In this way, by analyzing many trajectories computed from slightly different initial conditions, different distributions such as photoelectron spectra or the relative yield of different fragmentation channels can be calculated. In addition, if unexpected features are found in the so obtained spectra, the physical reason behind can be tracked down by analyzing the trajectories contributing to the feature under consideration.

The most difficult problem that has to be solved when constructing a model for classical laser-atom and laser-molecule interaction is the stability of the ground state (the state of lowest energy). In order for the model to be self-consistent, the atom or molecule should not spontaneously ionize or dissociate before the interaction with the laser field. In general, a collection of negatively and positively charged particles interacting with the Coulomb force only is not stable. The reason is that due to the electron-electron interaction, one electron can be ejected at the same time as another electron is pulled closer to the nucleus, with the total energy of the system still being conserved. For one-electron systems with a single nucleus (hydrogen atom and hydrogen-like ions), the Kepler orbits with a certain total energy provide a classical ground state [62–64], but we are interested in many-particle dynamics and consequently have to look for other solutions.

One commonly employed method is to modify the Coulomb potential at short interparticle distances, so that the potential energy takes a finite value even when two particles are located at the same place. The most popular choice is the so-called soft-core potential,

$$V_{sc}(\mathbf{r}_1, \mathbf{r}_2) = \frac{e^2 Z_1 Z_2}{\sqrt{(\mathbf{r}_1 - \mathbf{r}_2)^2 + \alpha^2}}, \tag{2.1}$$

where $e = |e|$ is the elementary charge, $Z_{1,2}$ are the charge numbers of the particles, and α is a constant. In (2.1), Gaussian units have been employed, as will be done throughout this chapter. A classical model of the helium atom may be created by letting two electron interact with each other and with a stationary nucleus via soft-core potentials. Including also the force from the electric field of the laser pulse, this kind of model has been used extensively to study NSDI in He [65–83], and also NSMI [78, 79, 84–86]. In [67], it was shown that the "knee" structure in the NSDI probability curve could be reproduced by a purely classical model, showing that the electron-electron correlation in the NSDI process can be understood classically. In fact, all three steps in the three-step model [9]: (1) electron ejection, (2) acceleration in the field, and (3) recollision, are classical processes, provided that the first step proceeds by over-the-barrier ionization. However, for sufficiently low laser intensity, the classical probability for electron ejection is zero, while the quantum mechanical probability of tunneling through the barrier may still be substantial [87]. Therefore, an extension of the purely classical soft-core model of He has been proposed, in which electrons are allowed to be ejected by quantum mechanical tunneling [7, 88–92]. The tunneling probability is given either by the ADK formula [93] or by a WKB-type approximation.

Another way to avoid autoionization, suggested first in the context of nuclear physics [94], is the following [95]: In addition to the usual Coulomb potential, a momentum-dependent potential is added to the Hamiltonian for the electron-nucleus interaction. In the case of a hydrogen atom, the Hamiltonian reads (assuming that the nucleus is fixed)

$$H_{\mathrm{H}} = \frac{\mathbf{p}^2}{2m} - \frac{Ze^2}{|\mathbf{r}|} + g(|\mathbf{p}|, |\mathbf{r}|), \tag{2.2}$$

where m is the mass of the electron, $\mathbf{r}$ is the position of the electron, $\mathbf{p}$ is the momentum, Z is the charge number of the nucleus, and g is a function that will be discussed below in Sect. 2.3. The function g effectively acts to keep the electron in the region of the phase space where $|\mathbf{r}||\mathbf{p}| \geq \hbar$ is satisfied, and can therefore be said to be a classical approximation to the Heisenberg principle.

The rest of this chapter is organized as follows. We continue in Sect. 2.2 with a general overview of the theoretical methods employed in classical trajectory Monte Carlo simulations on atomic and molecular systems. The following sections are devoted to examples from our recent research activities in this area. In Sect. 2.3, we introduce a particular form of classical trajectory model called "Fermionic Molecular Mechanics", and describe two applications: Recollision-induced core electron ejection in Sect. 2.3.1, and laser-driven dynamics of the $D_3{}^+$ molecular ion in Sect. 2.3.2. We continue in Sect. 2.4, which contains examples of many-particle systems modeled by purely classical mechanics: A soft-core model of $H_3{}^+$ (Sect. 2.4.1), laser-driven He (Sect. 2.4.2), and the proton recollision dynamics in ^{15}NH (Sect. 2.4.3). We conclude with a summary in Sect. 2.5.

Energies are sometimes stated in units of Hartree, $E_h \approx 27.2$ eV, and lengths in units of the Bohr radius $a_0 \approx 0.53$ Å.

2.2 Theoretical Methods

In this section we briefly recall the theoretical principles and numerical methods which the classical trajectory simulations are based on. The first step is to fix the model for the system under consideration. In classical mechanics, this means that the Hamiltonian H should be defined. In general, the Hamiltonian is a function of the positions $\mathbf{r}_1, \mathbf{r}_2, \ldots$ and the momenta $\mathbf{p}_1, \mathbf{p}_2, \ldots$ of all the particles. In the following discussion, we will use the term "configuration" to denote a specification of all position vectors and momentum vectors $\{\mathbf{r}_i, \mathbf{p}_i\}, i = 1, \ldots, N$ (a point in classical phase space) and H represents the total energy of the system. Having defined the Hamiltonian, the equations of motion to be solved are

$$\frac{d\mathbf{r}_i}{dt} = \frac{\partial H}{\partial \mathbf{p}_i}, \qquad \frac{d\mathbf{p}_i}{dt} = -\frac{\partial H}{\partial \mathbf{r}_i} + q_i \mathbf{E}(t, \mathbf{r}_i) + q_i \frac{d\mathbf{r}_i}{dt} \times \frac{\mathbf{B}(t, \mathbf{r}_i)}{c}, \qquad 1 \le i \le N. \tag{2.3}$$

Note that we have added the Lorentz force arising from the external laser field, proportional to the charge q_i of the particle i. In (2.3), c is the speed of light, and the electric field and the magnetic field of the laser are denoted by $\mathbf{E}(t, \mathbf{r})$ and $\mathbf{B}(t, \mathbf{r})$, respectively.

In all examples presented below, we have assumed the following form for the laser pulse,

$$\mathbf{E}(t, \mathbf{r}) = \hat{\mathbf{e}} E_0 \Phi(t - \hat{\mathbf{k}} \cdot \mathbf{r}/c), \qquad \mathbf{B}(t, \mathbf{r}) = \hat{\mathbf{k}} \times \mathbf{E}(t, \mathbf{r}), \tag{2.4}$$

$$\Phi(\tau) = \begin{cases} \sin(\omega_0 \tau + \varphi_0) \sin^2\left(\frac{\pi \tau}{T_0}\right) & \text{if } 0 \le \tau \le T_0, \\ 0 & \text{otherwise.} \end{cases} \tag{2.5}$$

Here $\hat{\mathbf{e}}$ is the polarization direction of the laser field and $\hat{\mathbf{k}}$ is the propagation direction. The peak field strength is denoted by E_0, the angular frequency by ω_0, the carrier-envelope phase (CEP) by φ_0, and the pulse length by $T_0 = 2\pi n_{\text{cyc}}/\omega_0$, where n_{cyc} is the total number of optical cycles in the laser pulse. We point out that

$$\int_0^{T_0} \Phi(\tau) d\tau = 0, \tag{2.6}$$

which is required for a propagating laser pulse. In some applications presented below, we will make the dipole approximation, and formally let $c \to \infty$. This results in an electric field independent of the position $\mathbf{r}_i$, and $\mathbf{B}/c = 0$. For the conversion to laser field intensity I_0, we use the formula $I_0/(\text{W cm}^{-2}) \approx 1.33 \times 10^{-7} [E_0/(\text{V m}^{-1})]^2$.

In the field-free situation ($E_0 = 0$), the total energy and the total momentum are conserved: a solution to (2.3) satisfies $\mathrm{d}H(t)/\mathrm{d}t = 0$ and $(\mathrm{d}/\mathrm{d}t) \sum_i \mathbf{p}_i(t) = \mathbf{0}$. Even when the laser field is present, due to the property (2.6) the momentum along the polarization direction $\hat{\mathbf{e}}$ is the same before and after the laser pulse, $\hat{\mathbf{e}} \cdot \sum_i \mathbf{p}_i(-\infty) = \hat{\mathbf{e}} \cdot \sum_i \mathbf{p}_i(\infty)$. Along the propagation direction $\hat{\mathbf{k}}$, there can be a change in momentum due to absorption of photons: We have $\mathrm{d}H/\mathrm{d}t = \sum_i q_i (\mathrm{d}\mathbf{r}_i/\mathrm{d}t) \cdot \mathbf{E}(t, \mathbf{r}_i) = c\hat{\mathbf{k}} \cdot (\sum_i \mathrm{d}\mathbf{p}_i/\mathrm{d}t)$, which means that $H(\infty) - H(-\infty) = c\hat{\mathbf{k}} \cdot \sum_i [\mathbf{p}_i(\infty) - \mathbf{p}_i(-\infty)]$.

Given initial values $\mathbf{r}_i^{(0)}, \mathbf{p}_i^{(0)}, i = 1, \ldots, N$, we can solve the equations of motion (2.3). Numerically, this is typically accomplished with a Runge-Kutta algorithm with a variable step size (see for example the implementation in [96]). As a result of the interaction with the laser field, energy is absorbed by the system, leading to ionization, dissociation, or both. Therefore, we need to integrate the equations of motion at least until the laser pulse vanishes at $t = T_0$, but in many cases a longer integration time is needed in order to securely determine the final state of the system. Whether an electron is ionized or not is judged by its single-particle energy, and a positive energy signifies an ejected electron. Different fragmentation channels can be distinguished by comparing the internuclear distances.

A simulation is conducted according to the following steps: First, the initial conditions (momenta and positions) of the system are sampled according to a predefined probability distribution. Usually the probability distribution is derived from a microcanonical ensemble, meaning that all initial states have the same energy. Second, the time-dependent trajectory of the system is obtained by numerically solving the equations of motion (2.3). Finally, the final state of the system (dissociated and/or ionized and/or excited) is checked and recorded. These steps are repeated many times. Typically 10^4–10^5 trajectories are run at each value of the laser parameters, but sometimes more trajectories are needed if a pathway with a low relative yield is studied. Even though many trajectories may be required, each trajectory is calculated independently of the results of other trajectories. This means that the simulation procedure can easily be parallelized to run on many computer processors simultaneously, for example at a computer cluster, which is important to acquire a large enough number of trajectories for processes with small relative yield.

After the accumulation of many trajectories run with slightly different initial positions, quantities corresponding to experimental observables, such as photoelectron energy distributions, or spectra of the released kinetic energy, are obtained by counting the number of trajectories in a particular final channel with an energy or momentum in a particular range. One of the strong features of classical trajectory methods is the possibility of selective analysis of trajectories leading to a specific final channel. For example, in NSDI, we may isolate only the trajectories leading to double ionization, and statistically analyze these trajectories at different points in time [66]. This kind of analysis can lead to a deeper understanding of many types of laser-induced many-particle dynamics.

2.3 Fermionic Molecular Dynamics

Fermionic Molecular Dynamics (FMD) originates from the following idea [94, 95]: Since the origin of the instability of a classical, many-electron atom is the possibility for an electron to collapse into the nuclear core, we prevent this possibility by adding a suitable potential term to the Hamiltonian. However, only slow electrons need to be prevented from approaching close to the nucleus, fast electrons (for example scattering electrons, or core electrons in the case of a heavy atom) can be allowed to have a small electron-nucleus distance. This reasoning leads us to consider potentials that implement the inequality $|\mathbf{r}||\mathbf{p}| \geq \xi_0\hbar$ with ξ_0 being a constant of the order of one.

This kind of constraint may be thought of as a way of introducing the Heisenberg principle in classical mechanics. In the same way, we assign a spin label to each electron (up or down), and introduce potentials which prevent two electrons with the same spin from occupying the same point in classical phase space. This corresponds to a classical form of the Pauli exclusion principle. The original form of the FMD Hamiltonian for N_e electrons and N_n nuclei reads [95]

$$H_{\mathrm{FMD}} = \sum_{j=1}^{N_n} \frac{\mathbf{P}_j^2}{2M_j} + \sum_{i=1}^{N_e} \frac{\mathbf{p}_i^2}{2m} + \sum_{j=1}^{N_n}\sum_{i=1}^{N_e}\left[-\frac{e^2 Z_j}{\rho_{ij}} + \frac{\hbar^2}{\mu_j \rho_{ij}^2} f(Q_{ij}, \rho_{ij}, \xi_0)\right]$$
$$+\sum_{i=1}^{N_n}\sum_{j>i} \frac{e^2 Z_i Z_j}{R_{ij}} + \sum_{i=1}^{N_e}\sum_{j>i}\left[\frac{e^2}{r_{ij}} + \delta_{\sigma_i \sigma_j} \frac{2\hbar^2}{m r_{ij}^2} f(q_{ij}, r_{ij}, \eta_0)\right]. \tag{2.7}$$

In (2.7) we have used $\mathbf{r}_i$, $\mathbf{p}_i$, and σ_i ($=\uparrow$ or $\downarrow$) to denote the position, momentum and spin of electron i, $\mathbf{R}_j$, $\mathbf{P}_j$, Z_j, and M_j to denote the position, momentum, charge number and mass of nucleus j, and $\mu_i = mM_i/(M_i + m)$ is a reduced mass. Moreover, we use the following definitions for the relative momenta and coordinates, $Q_{ij} = |m\mathbf{P}_j - M_j\mathbf{p}_i|/(M_j + m)$, $\rho_{ij} = |\mathbf{r}_i - \mathbf{R}_j|$, $R_{ij} = |\mathbf{R}_i - \mathbf{R}_j|$ $q_{ij} = |\mathbf{p}_i - \mathbf{p}_j|/2$, $r_{ij} = |\mathbf{r}_i - \mathbf{r}_j|$. The function f is given by [95, 97]

$$f(q, r, \xi) = \frac{\xi^2}{16} \exp\left\{4\left[1 - (qr/\xi\hbar)^4\right]\right\}. \tag{2.8}$$

We take $\xi_0 = 1/\sqrt{1 + 1/8} \approx 0.94$ and $\eta_0 = 2.767/\sqrt{1 + 1/8} \approx 2.61$ as the numerical values of the constants ξ_0 and η_0 [95]. The Hamiltonian (2.7) has been shown to give stable classical atoms with $Z \leq 94$ [98, 99], and also stable, classical H_2 and $H_2{}^+$ molecules [97]. The ground state, defined as the configuration which minimizes the total energy H_{FMD}, can be found numerically by employing a suitable minimization algorithm. We used the downhill simplex method described in [96, 100]. In the ground state, the particle velocities are zero, $\mathrm{d}\mathbf{r}_i/\mathrm{d}t = \mathrm{d}\mathbf{R}_i/\mathrm{d}t = \mathbf{0}$, but the momenta are non-vanishing, $\mathbf{P}_i = \mathbf{p}_i \neq \mathbf{0}$. Typically, the momentum for a valence electron in the ground state is $|\mathbf{p}| \approx \hbar/a_0$ [95, 98]. We point out that a configuration that starts out in the ground state will stay there for all times in the absence of a laser field.

An important symmetry property satisfied by H_{FMD} is that applying a rotation to all momentum vectors or a rotation to all position vectors gives the same energy. The FMD model has been used to simulate atomic and molecular collision processes [97, 101–108], and laser-driven processes in atoms [109–115], molecules [116–119], and Xe clusters [120]. A more detailed account of the FMD model can be found in the reviews [121, 122].

2.3.1 *Inner-Shell Electron Ejection in a Laser-Driven Carbon Atom*

Although there exist a large body of literature describing experiments and calculations on NSDI in two-electron systems (or systems which can be approximated as two electron systems), comparatively little is known about non-sequential multiple ionization (NSMI) [10–19]. With NSMI, we have in mind the process where one electron is ejected by field ionization, and later recollides with the atomic core to eject two or more bound electrons, resulting in the formation of a highly charged final state. The major topic in the theoretical calculations has so far been the non-sequential triple ionization process [78, 79, 84–86].

In this section, we summarize our recent investigation of NSMI of a laser-driven carbon atom [123]. The carbon atom is described by the standard FMD Hamiltonian (2.7), with $N_n = 1$ and $N_e = 6$, assuming three up-spin and three down-spin electrons. Interestingly, the inclusion of the spin-spin interaction term [the term proportional to $\delta_{\sigma_i\sigma_j}$ in (2.7)] leads to an electronic shell structure of the classical ground state [98], with two electrons in each shell. In the configuration with lowest energy, some electrons are situated close to the nucleus and may be labeled as "core electrons", while other, more loosely bound electrons can be categorized as valence or inner valence. The distinction between the different types of electrons can be made quantitatively by defining a single-particle energy ε_i as

$$\varepsilon_i = \frac{\mathbf{p}_i^2}{2m} + \sum_{j=1}^{N_n} \left[-\frac{e^2 Z_j}{\rho_{ij}} + \frac{\hbar^2}{\mu_j \rho_{ij}^2} f(Q_{ij}, \rho_{ij}, \xi_0) \right] + \frac{1}{2} \sum_{j \neq i} \left[\frac{e^2}{r_{ij}} + \delta_{\sigma_i\sigma_j} \frac{2\hbar^2}{m r_{ij}^2} f(q_{ij}, r_{ij}, \eta_0) \right], \tag{2.9}$$

assigned to each electron [109, 123]. In the ground state, an electron with a large negative value of ε_i indicates a strongly bound core electron, while the valence electrons have smaller (negative) values.

The shell structure of the FMD ground state enables us to discuss the ejection of core electrons by recolliding electrons. Core, or inner-shell electron ejection induced by recollision is a subject that has not been fully exploited, although a few experimental reports on the existence of the process have been published [124, 125]. For example, one might speculate in the creation of core-hole atoms controlled by the CEP of the laser pulse, considering that the probability of recollision and the kinetic energy of the recolliding electron are highly sensitive to the CEP [126, 127].

In Fig. 2.1a, we show the probability of the ionization into a particular final charge state of the C atom as a function of the incident laser field intensity. The laser parameters used are $\omega_0 = 0.057 E_h/\hbar$ (corresponding to a wavelength of 800 nm), $n_{\text{cyc}} = 3$, and $\varphi_0 = \pi/2$. In the simulation approximately 3×10^5 trajectories were calculated at each laser field intensity. Both the electric and the magnetic field

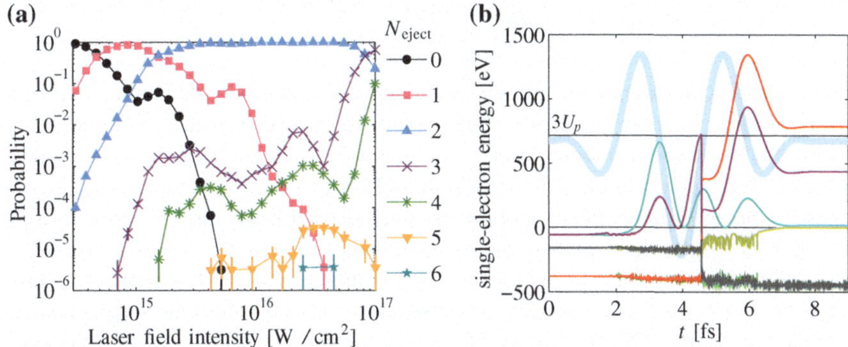

Fig. 2.1 **a** Ionization probability with the ejection of N_{eject} electrons in the final state as a function of laser field intensity. *Statistical error bars* are indicated when they exceed the size of the symbols. **b** Example of a trajectory leading to the ejection of four electrons, plotted as the single-electron energy ε_i [see (2.9)] versus time. The laser field intensity is $I_0 = 4 \times 10^{15}$ W/cm^2. Different *line colors* correspond to different electrons. The electric field of the laser pulse (on an arbitrary scale) is shown in the background as a *thick line*. The maximum energy at recollision, that is, the ponderomotive energy times three, $3U_p = 3e^2E_0^2/(4m\omega_0^2) \approx 716$ eV is indicated as a *horizontal line*

contributions to the Lorentz force were included in the equations of motion. The initial values for the simulations were generated by applying a random rotation to the position vectors and the momentum vectors of the ground state configuration. As the FMD Hamiltonian (2.7) is invariant under such rotations, different ground state configurations with the same energy can be generated in this way.

As can be seen in Fig. 2.1a, we are able to simulate NSMI of a laser-driven carbon atom. Clear plateau structures can be seen in the channels involving the ejection of 3 and 4 electrons. In the plateau region (3×10^{15} W/cm$^2 \le I_0 \le 3 \times 10^{16}$ W/cm^2), all trajectories of these channels involve NSMI: one ejected valence electron returns to knock out two bound electrons. An example of a NSMI trajectory, involving a rather complex multi-electron rearrangement process, is shown in Fig. 2.1b. First, the two valence electrons are ejected by the field ionization at around $t = 3$ fs. Then, at $t \approx 4.5$ fs, recollision occurs. The recolliding electron transfers a large amount of energy to two other electrons: one core electron, which is promptly ejected from the atomic core, and one inner valence electron which is excited first, and then ejected later (at $t \approx 7$ fs) after having absorbed additional energy from the laser field. It is interesting that this kind of electron rearrangement among electrons in different atomic shells, mediated by the electron-electron interaction, occurs in the classical trajectories.

2.3.2 $D_3{}^+$ in an Intense Laser Field

The $H_3{}^+$ molecular ion is the simplest stable polyatomic molecule, and is therefore of fundamental interest. In order to understand the laser-driven motion of complex many-body systems in intense laser fields beyond atoms and diatomic molecules, triatomic $H_3{}^+$ can be regarded as an ideal model system. Although pioneering experiments on the dynamics of $H_3{}^+$ and $D_3{}^+$ in intense laser fields have been performed [23–25, 128–130], no complete quantum mechanical simulation including both electronic and nuclear motion has been reported. Calculations on $H_3{}^+$ with stationary nuclei [131–135] do exist, as well as attempts to describe the electrons quantum mechanically with TDDFT with the three protons treated classically via Ehrenfest dynamics [136]. In experiments like [23–25, 129, 130], where a molecule is exposed to an intense laser pulse, many different ionization and dissociation channels are observed. The relative yield of each channel depends on the laser parameters, such as intensity, wavelength, and pulse length. One of the ultimate goals for theory is to reproduce the measured variations in the relative yields as a function of the laser parameters, which is, however, extremely difficult.

Motivated by the experiments reported in [23, 128], we have made an attempt to simulate and understand the laser-driven dynamics of $D_3{}^+$ [117, 119]. The model used is the FMD Hamiltonian (2.7), augmented with certain additional three- and four body potentials first suggested by Cohen [97], and a five-body potential introduced by us. These additional potentials are introduced to reproduce the quantum mechanical potential energy curves. The original FMD Hamiltonian (2.7) suggested in [95] gives too large values for the binding energies of H_2, $H_2{}^+$ [95] and $H_3{}^+$ [119]. It is interesting to note, however, that although the binding energies are too large, the original FMD Hamiltonian (2.7) *does* produce stable ground states for both H_2 and $H_2{}^+$ [97]. We have also checked that the original FMD Hamiltonian (2.7) yields a stable structure of $H_3{}^+$.

The Hamiltonian used for the $D_3{}^+$ simulations was given by

$$\begin{aligned} H_{D_3^+} &= H_{\mathrm{FMD}}(N_e = 2,\ N_n = 3, M_i = M_{\mathrm{D}}) \\ &+ \sum_{j=1}^{3}\sum_{k>j}\sum_{l=1}^{2} \frac{\hbar^2}{\nu R_{jk}^2} f(Q_{jkl}, S_{jkl}, \chi_1) \\ &+ \sum_{j=1}^{3}\sum_{k>j} \frac{2\hbar^2}{m R_{jk}^2} f(q_{12}, r_{12}, \chi_2) - \frac{\hbar^2 f(q_{12}, r_{12}, \chi_3)}{m(R_{12}+R_{13}+R_{23}+a_0/10)^2}, \end{aligned} \tag{2.10}$$

where M_{D} is the mass of the deuteron, $\chi_1 = 0.90$ [97], $\chi_2 = 1.73$ [97], $\chi_3 = 1.85$, $Q_{jkl} = |m(\mathbf{P}_j + \mathbf{P}_k)/2 - 2M_{\mathrm{D}}\mathbf{p}_l|/(2M_{\mathrm{D}} + m)$, $S_{jkl} = |\mathbf{R}_j/2 + \mathbf{R}_k/2 - \mathbf{r}_l|$, and $\nu = 2mM_{\mathrm{D}}/(2M_{\mathrm{D}} + m)$.

The lowest energy $\varepsilon_{\mathrm{min}}^{D_3^+} = -1.34E_h$ of the Hamiltonian (2.10) is obtained when the three nuclei form an isosceles triangle with side lengths $1.46a_0$, $1.46a_0$, and $1.55a_0$.

Hereafter, this molecular geometry, together with the coordinates and momenta of the electrons giving the lowest energy, is called the ground state configuration. We recall that a configuration is here defined as a specification of the momentum vectors $\{\mathbf{P}_j, \mathbf{p}_i\}$ and the coordinate vectors $\{\mathbf{R}_j, \mathbf{r}_i\}$ of all the particles. The energy $\varepsilon_{\rm min}^{\rm D_3^+}$ should be compared with the results from accurate quantum chemical (QC) calculations [137]: the minimum energy $\varepsilon_{\rm min}^{\rm D_3^+}(\rm QC) = -1.34E_h$ and an equilibrium molecular structure of an equilateral triangle with side lengths $1.65a_0$. The existence of a global minimum of the Hamiltonian (2.10) (the ground state) implies that in the absence of perturbations, a trajectory starting in the ground state will stay in the ground state forever, and will not dissociate or ionize. Also the ground states of D_2 [defined by ignoring terms in (2.10) involving $\mathbf{R}_3$ and $\mathbf{P}_3$] and $D_2{}^+$ [defined by ignoring terms in (2.10) involving $\mathbf{R}_3$, $\mathbf{P}_3$, $\mathbf{p}_2$ and $\mathbf{r}_2$] are stable in the same sense.

For the time-dependent simulations, the nonrelativistic approximation of the equation of motion (2.3) with $c = \infty$ was employed. In order to mimic the experimental situation in [23], the initial values were sampled from a distribution where an additional vibrational energy $\varepsilon_{\rm vib} = (0.08 \pm 0.02)E_h$ was added to the ground state configuration. In practice, initial configurations were selected by displacing first the three deuterons randomly from the equilibrium configuration, minimizing the Hamiltonian with respect to the two electrons, and keeping only those configurations which have a total energy $\varepsilon_{\rm tot}$ satisfying $0.06E_h \le \varepsilon_{\rm tot} - \varepsilon_{\rm min}^{\rm D_3^+} \le 0.1E_h$. We remark that the minimum energy of the H_2 molecule in our model is $\varepsilon_{\rm min}^{\rm H_2} \approx -1.17E_h$ [97]. This means that a $D_3{}^+$ molecule vibrationally excited by $\varepsilon_{\rm vib} = 0.1E_h$ will not dissociate to $D_2 + D^+$ in the absence of a laser field, since $\varepsilon_{\rm min}^{\rm D_3^+} = -1.34E_h < \varepsilon_{\rm min}^{\rm H_2} - \varepsilon_{\rm vib}$. In addition, a random rotation is applied to the coordinate vectors of the electrons and deuterons in the initial configuration, simulating a molecule in the rotational ground state.

In Fig. 2.2a, we show results of Monte Carlo simulations performed at different peak field intensities. Around $N_{\rm tot} = 10^4$ trajectories were run at each value of the laser intensity. The other laser field parameters were $\omega_0 = 0.058E_h/\hbar$ (wavelength $\lambda = 790$ nm), $n_{\rm cyc} = 3$, and $\varphi_0 = 0$. Remarkably, all possible fragmentation channels are produced in the simulation. All the possible particle rearrangement processes are identified, which may not always be realized by a classical model of a polyatomic molecule. It is possible that a certain model gives a limited set of dissociation/ionization processes only, other quantum mechanically possible pathways are closed due to the structure of the Hamiltonian.

In Fig. 2.2b we show the calculated spectrum of the released kinetic energy. The released kinetic energy $\mathscr{E}_{\rm kin}$ is defined as the sum of the kinetic energies of all heavy fragments in the final state (the kinetic energies of the electrons are not included). Compared to the experimentally measured spectra of the released kinetic energy (see Fig. 2 in [23]), a good agreement can be seen qualitatively. The widths and the relative positions of the simulated curves for the three ionization channels are similar to the experimental curves. The simulation does not, however, reproduce the energy scale quantitatively, that is, the simulated $\mathscr{E}_{\rm kin}$ is consistently too high by a factor of

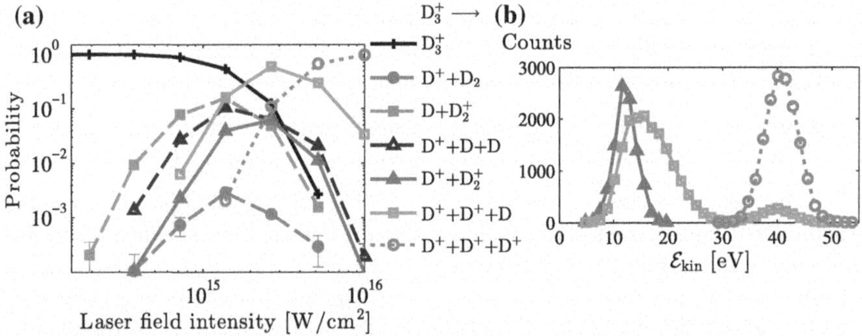

Fig. 2.2 **a** Total probability as a function of laser field intensity for each of the possible final channels. *Statistical error bars* are indicated when they exceed the size of the *curve symbols*. **b** Spectra of the released kinetic energy $\mathscr{E}_{\text{kin}}$ at a laser field intensity of $I_0 = 4 \times 10^{15}$ W/cm^2. The *curve* for the $D_3{}^+ \to D^+ + D_2{}^+$ channel (*filled triangles*) was multiplied by a factor of 10

roughly 1.5. The reason for this disagreement is that the model Hamiltonian (2.10) results in a too "stiff" $H_3{}^+$ molecule [119], so that after the ejection of one electron, the Coulomb explosion starts at a too small internuclear distance on average.

More interesting than a quantitative comparison is the physical insight that one gains from the classical simulation. We showed in [117] that the reason for the different widths of the $\mathscr{E}_{\text{kin}}$ curves for the two single ionization channels (the width of the $D_3{}^+ \to D^+ + D_2{}^+$ channel being narrower than that of the $D_3{}^+ \to D^+ + D^+ +$ D channel) is that the $D_3{}^+ \to D^+ + D^+ +$ D channel allows for a higher excitation of the bound electron, which provides less screening for the nuclear repulsion. We also pointed out the interesting possibility of incomplete electron ejection [138], which manifests itself as a small peak in the $D_3{}^+ \to D^+ + D^+ +$ D channel having the same value of the released kinetic energy (of around 40 eV, see Fig. 2.2b) as the $D_3{}^+ \to D^+ + D^+ + D^+$ channel. This peak originates from an electron that is ejected, but later recaptured into a loosely bound Rydberg state. In $D_3{}^+$, this process was recently confirmed experimentally [129].

2.4 Purely Classical Models

Purely classical models are models where all particles involved are treated as classical point particles, and where the interactions via two-body potentials depend only on the relative interparticle distances. No "ad hoc" potentials (like in the FMD method described in Sect. 2.3) are added to the Hamiltonian. In addition to the interparticle forces, we also add the forces induced by the external laser field in the equations of motion. As already mentioned in the introduction, this kind of approach necessitates the use of modified interaction potentials. In the two first examples described in this section, the soft-core potential (2.1) is employed.

2.4.1 Soft-Core Model of $H_3{}^+$

Motivated by the same arguments as presented in Sect. 2.3.2, we are interested in constructing a classical model of the $H_3{}^+$ molecular ion to study its response to an intense laser field. In this section, we review our recent approach to this problem, using soft-core Coulomb potentials to remedy the inherent instability of classical trajectory models [139]. The same approach has been applied before to model diatomic molecules [140–142], but (apart from our paper [139]) not to polyatomic molecules.

The Hamiltonian for the soft-core $H_3{}^+$ system is defined as

$$H^{sc}_{\mathrm{H}_3^+} = \sum_{j=1}^{3} \frac{\mathbf{P}_j^2}{2M} + \sum_{k=1}^{2} \frac{\mathbf{p}_k^2}{2m} + V^{sc}_{\mathrm{H}_3^+}, \tag{2.11}$$

$$V^{sc}_{\mathrm{H}_3^+} = \frac{e^2}{\sqrt{(\mathbf{r}_1 - \mathbf{r}_2)^2 + \alpha_{ee}^2}} + \sum_{j=1}^{3} \sum_{k>j} \frac{e^2}{\sqrt{(\mathbf{R}_j - \mathbf{R}_k)^2 + \alpha_{pp}^2}} - \sum_{j=1}^{3} \sum_{k=1}^{2} \frac{e^2}{\sqrt{(\mathbf{R}_j - \mathbf{r}_k)^2 + \alpha_{ep}^2}}, \tag{2.12}$$

where M is the mass of the proton, and α_{ee}, α_{ep}, and α_{pp} are three soft-core parameters related to the electron-electron interaction, the electron-proton interaction, and the proton-proton interaction, respectively. The numerical values of α_{ee}, α_{ep} and α_{pp} are not arbitrary. Only certain choices will lead to stable classical ground states of the $H_3{}^+$ molecular ion. We also require that $H_2{}^+$ and H_2 are stable, and that $H_3{}^{2+}$ is unstable. By "stable", we mean in the current context that there is a particle configuration (with $\mathbf{P}_j = \mathbf{p}_k = \mathbf{0}$ for all j and k) which minimizes the total energy, and which is stable when perturbed slightly. In other words, the potential energy $V^{sc}_{\mathrm{H}_3^+}$ should have a global minimum, as should the corresponding potentials energies for H_2 and $H_2{}^+$. If the equations of motion (2.3) are solved with $E_0 = 0$, a particle configuration starting at a global minimum will stay there for all time. This notion of a ground state is the same as for the FMD model in Sect. 2.3 in the sense that all particles are stationary. The difference is that in the purely classical case, the particle momenta are zero. The acceptable sets of numerical values for α_{ee}, α_{ep} and α_{pp} were systematically investigated in [139]. For the examples we show below, $\alpha_{ee} = 1.4991a_0$, $\alpha_{ep} = 1.1560a_0$, and $\alpha_{pp} = 0.8669a_0$ were adopted.

We introduce here a few examples of the results of the time-dependent simulations. The non-relativistic approximation with $c = \infty$ was used with $\omega_0 = 0.058E_h/\hbar$, $n_{\mathrm{cyc}} = 3$, and the CEP $\varphi_0 = 0$. Similarly to the procedure described in Sect. 2.3.2, initial values for the simulation runs were obtained by taking points on a trajectory slightly excited compared to the ground state configuration, which was calculated by adding an amount of $0.03E_h$ of energy to the ground state of the system. Then, we solve the equations of motion in the absence of the laser field. A random rotation

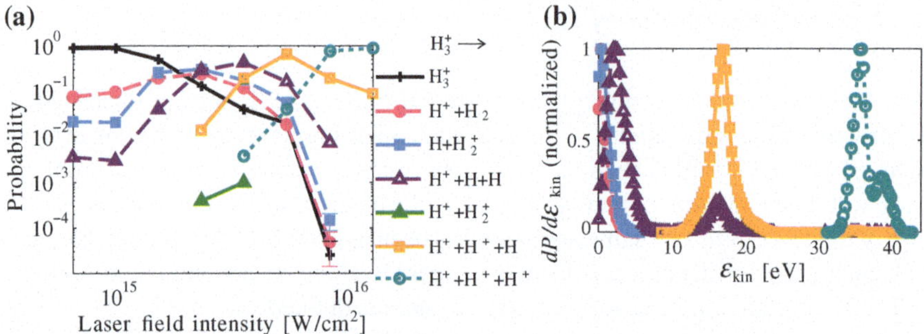

Fig. 2.3 **a** Total probability as a function of laser field intensity for each of the possible final channels. The *statistical error bars* are shown when they are larger than the *symbols*. **b** Normalized spectra of the released kinetic energy $\mathcal{E}_{\text{kin}}$. The laser field intensity is $I_0 = 5 \times 10^{15}$ W/cm^2

of the coordinate vectors of all particles was performed before starting each run, in order to simulate a randomly aligned molecular ensemble. In Fig. 2.3a, we show the total probabilities of the possible dissociation and ionization channels, as a function of the laser field intensity. Around 4×10^4 trajectories were calculated at each value of the laser field intensity. We can see that all possible dissociation and/or ionization channels appear as a result of the simulation. However, the $H_3{}^+ \rightarrow H^+ + H_2{}^+ + e^-$ channel is only produced at two values of the laser field intensity, with a negligibly small number of events. This disagrees with both experimental measurements [23] and the FMD model of laser-driven $D_3{}^+$ [117]. Also the simulated spectra of the released kinetic energy shown in Fig. 2.3b fail to reproduce the results in [23]. We can say that the introduction of the auxiliary potentials in the FMD model in Sect. 2.3 play an essential role in reproducing the experimental data we considered. We conclude that the soft-core model of $H_3{}^+$ cannot be employed to interpret experimental spectra, but could be used to study theoretically the laser-induced energy sharing among electrons and protons and other strong-field-driven many-body phenomena in classical mechanics.

2.4.2 Magnetic Field Effects in Non-sequential Double Ionization

At large field intensities and/or long wavelengths, the motion of a free electron in the laser field departs from an oscillatory motion along the laser polarization direction [143]. Due to the $(\mathbf{v}/c) \times \mathbf{B}$ part of the Lorentz force, the electron also moves in the propagation direction of the laser pulse, a direction perpendicular to both the **E** and the **B** field. For a monochromatic laser field, the distance traveled in the propagation direction during one field cycle is $\delta z = 2\pi U_p/(cm\omega_0)$, where $U_p = e^2 E_0^2/(4m\omega_0^2)$ is the ponderomotive potential. From this formula, we see that the motion along the propagation direction is an effect of relativistic origin (scaling with $1/c$). The $1/\omega_0^3$ scaling indicates that the effect is expected to become more important when the wavelength is increased. The drift motion of the free electron along the propagation direction decreases the recollision probability: at sufficiently high laser field intensities and/or sufficiently long wavelengths, the electron cannot return to the atomic core after the ejection. In order that the electron comes back to the core even at high laser field intensity, several ideas for laser field configurations were proposed, such as two counterpropagating, linearly polarized laser pulses with a time delay [144], or two counterpropagating circularly polarized laser pulses [145].

At high velocities of the electron, there is another relativistic effect that may become important in a multi-electron system: the magnetic field component of the electron-electron interaction. If we consider the motion of one electron as seen in the rest frame of another electron, the electron at rest experiences a time-dependent electric field due to the repulsive Coulomb potential of the moving electron. From Maxwell's equations, we know that a time-dependent electric field is always accompanied by a magnetic field. The effect of this magnetic field can be described as a correction to the equations of motion proportional to $1/c^2$ [146]. The explicit form for the equations of motion can be derived from the Darwin Lagrangian [147].

The purpose of our recent investigation [81], which we summarize in this section, was to answer the following two questions: (1) How important is the magnetic part of the Lorentz force for the NSDI process? (2) How important are the $1/c^2$ corrections to the equations of motion arising from the magnetic field part of the electron-electron interaction in the NSDI process? The theoretical model we introduced is classical He: two electrons moving in the soft-core potential (see (2.1)) of a nucleus with $Z = 2$ and soft-core parameter $\alpha = 0.79a_0$. A similar model of classical He has been employed frequently in the past, and is still used to study different aspects of the non-relativistic NSDI process [65–80, 82, 83]. The simulations were carried out at three different levels of approximation: (i) Non-relativistic approximation. Here $c \to \infty$, which means that there are no terms depending on the speed of light in the equations of motion and no Lorentz force originating from the magnetic field. (ii) Non-relativistic approximation with magnetic field. Here the $(\mathbf{v}/c) \times \mathbf{B}$ part of the Lorentz force is included in the equations of motion. (iii) Darwin approximation. Both magnetic field terms of order $1/c$ and Darwin corrections of order $1/c^2$ are included in the equations of motion. Initial values for the simulation were obtained from a microcanonical ensemble generated from points on a trajectory with a total energy equal to the quantum mechanical total energy of He.

In short, the results of our simulations showed that the Darwin corrections to the equations of motion proportional to $1/c^2$ were unimportant for the recollision process [81]. Since the Darwin corrections include terms $\propto (c|\mathbf{r}_1 - \mathbf{r}_2|)^{-2}$, one could in principle expect some contribution at a recollision event when the interelectronic distance $|\mathbf{r}_1 - \mathbf{r}_2|$ becomes small. However, in practice this effect was negligible: Double ionization probabilities calculated in the approximation (iii) were found to be equal to those calculated in the approximation (ii) within the statistical error bars.

On the other hand, the magnetic part of the Lorentz force plays an important role in the recollision process even at laser intensities as low as 10^{15} W/cm^2. Due to the

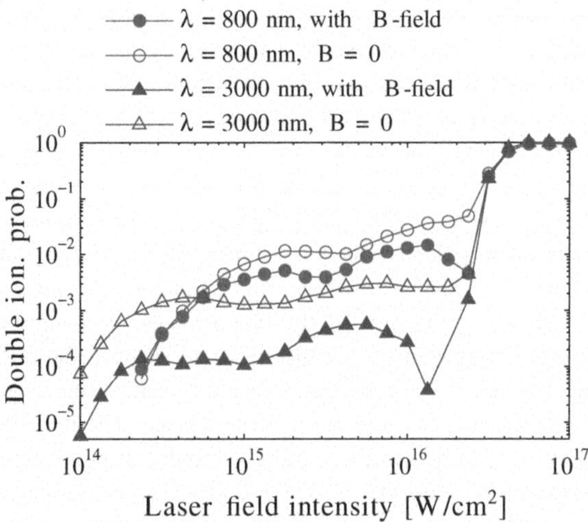

Fig. 2.4 Total probability for double ionization of He as a function of the laser field intensity, for two different wavelengths. Results obtained from simulations including (*solid symbols*) and excluding (*open symbols*) the **B** field are shown. The CEP used is $\varphi_0 = \pi/2$, and $n_{\text{cyc}} = 3$

force exerted on the electron in the propagation direction of the laser pulse, some trajectories (where the electron would have recollided in absence of the $(\mathbf{v}/c) \times \mathbf{B}$ force) coming back to vicinity of the atomic core fail to collide with the bound electron. The long trajectories, which require a full laser cycle to return to the nucleus, are particularly sensitive to the $(\mathbf{v}/c) \times \mathbf{B}$ force. Recollision still occurs for the short trajectories, in which there is approximately a half laser cycle between ejection and recollision.

In Fig. 2.4, we show results from a Monte Carlo simulation of a soft-core model of He exposed to a few-cycle laser pulse with two different wavelengths, $\lambda = 800$ nm and $\lambda = 3000$ nm. Approximately 10^6 trajectories were run at each value of the laser field intensity. The plateau structure in the double ionization probability versus intensity plot, characteristic of the NSDI process, is clearly visible. Interestingly, the effect of including the **B** field is much larger for the longer wavelengths: When $\lambda =$ 800 nm, the double ionization probability in the plateau region for the case with $\mathbf{B} = \mathbf{0}$ is around a factor of 2 larger than the double ionization probability in the case where the **B** field is included. For $\lambda = 3000$ nm, the double ionization probability with $\mathbf{B} = \mathbf{0}$ is one order of magnitude larger than in the case including the **B** field. Considering the recent trend of employing extremely long-wavelength (several μm) laser pulses for the generation of high-order harmonics [148–151], it would be worthwhile to investigate to what extent non-dipole effects are important in the high-order harmonic generation process [152]. The fact that relativistic effects become more important for long wavelengths has been stressed by Reiss [153]. We furthermore note that the investigation of atomic and molecular systems in long-wavelength laser fields by numerically solving the Schrödinger equation is problematic, since extremely large grid sizes are needed. The extension of the grid along the laser polarization direction has to be larger than at least the quiver amplitude $z_{\text{quiver}} = eE_0/(m\omega_0^2)$, which becomes $z_{\text{quiver}} \approx 2.3 \times 10^3\, a_0 \approx 0.1$ μm when the laser parameters are those used in Fig. 2.4 ($\lambda = 3000$ nm, $I_0 = 10^{16}$ W/cm^2). Classical trajectory methods do not have any problem with the large excursion of the ejected electrons, since the coordinates of the electron can be allowed to take any value.

2.4.3 Nuclear Reaction Induced by Proton Recollision

In this section, we demonstrate how classical trajectory Monte Carlo simulations can be used to estimate the reaction rates of a rather unusual process: The nuclear reaction $p + {}^{15}\mathrm{N} \rightarrow {}^{12}\mathrm{C} + \alpha$ triggered by proton recollision in an intense laser field [154]. The idea to employ laser fields to induce recollision with heavy particles to trigger nuclear reactions [155–161] is a natural extension of the highly successful concept of electron recollision. In order to estimate the required laser intensity, we note that the average quiver energy in a laser field (the ponderomotive potential) U_p is inversely proportional to the mass M of the accelerated particle, $U_p = q^2 E_0^2/(4M\omega_0^2)$, where q is the particle charge. For a proton, a quiver energy of 1 MeV, which would be required to initiate a nuclear reaction, is achieved at a laser intensity of $I_0 = 3 \times 10^{22}$ W/cm^2 (at a wavelength of 800 nm). This value of the laser intensity may sound extremely high, but laser pulses with a peak intensity of more than 10^{22} W/cm^2 have in fact been demonstrated more than 5 years ago [162].

In order to realize proton recollision in practice, we suggest to irradiate a small molecule with an intense laser pulse [154], as described in the following. In a first step, a small molecule (in the gas phase) containing a heavy nucleus and at least one proton is irradiated by an extremely strong laser pulse (intensity larger than 10^{22} W/cm^2). At the rising edge of the pulse, *all* electrons are ejected by direct field ionization, leaving a bare molecule consisting of positively charged ions. In absence of a driving field, the so created molecular ion would immediately dissociate via Coulomb explosion. In our case, however, the force exerted on the ions by the intense laser pulse can overcome the repulsive Coulomb force. A proton is subsequently accelerated by the laser field so as to recollide with a nearby, heavy particle at high relative kinetic energy, inducing a nuclear reaction. To be explicit, we consider the ^{15}NH molecule, since the $p + {}^{15}\mathrm{N} \rightarrow {}^{12}\mathrm{C} + \alpha$ reaction has a large cross section (about 0.1 barn at an impact energy of 0.5 MeV [163]).

The computational method follows the steps outlined in Sect. 2.2. The coupled equations of motion are solved for the proton and the $^{15}\mathrm{N}^{7+}$ nucleus, including the Coulomb repulsion, and the Lorentz force from the laser field (including both $\mathbf{E}$ and $\mathbf{B}$ fields). There is no need to use a soft-core potential, since we are not interested in bound state dynamics. The initial position of the proton relative to the ^{15}N nucleus is sampled from a distribution defined by the ground state vibrational wave function of the neutral ^{15}NH molecule and the degree of orientation of the molecular axis. Orientation of a molecular gas can be achieved with a weaker prepulse [164]. The degree of orientation is measured by the average of $\cos\theta$ [165], where θ is the angle between the molecular axis and the z axis. We will see that a strong orientation of the ^{15}NH molecule is advantageous for the recollision process, due to the $(\mathbf{v}/c) \times \mathbf{B}$ force of the laser field.

Differently from the other examples of classical trajectory calculations introduced in this chapter, the actual reaction (creation of a ^{12}C nucleus and an α particle) is not included in the trajectory model itself. This should be compared with the case of core electron ejection described in Sect. 2.3.1, where both the excursion of the free electrons in the continuum and the collision leading to the ejection of a core electron were included within the same theoretical model, without the need to introduce additional information about the probability for electron impact ionization. In the current case, since we are dealing with nuclear reactions occurring at a much smaller spatial scale, we include the reaction probability by adopting the known cross section σ [163] for the $p + {}^{15}\mathrm{N} \rightarrow {}^{12}\mathrm{C} + \alpha$ process. In practice, we estimate the recolliding flux of protons by counting all proton trajectories which pass the ^{15}N nucleus at a distance smaller than $b_0 = 0.05$ Å and have a relative kinetic energy larger than the effective threshold of 0.1 MeV [163] for the $p + {}^{15}\mathrm{N} \rightarrow {}^{12}\mathrm{C} + \alpha$ process. The reaction probability P is estimated by calculating

$$P = \frac{1}{n} \sum_{j=1}^{n} \frac{\sigma(\varepsilon_j^{(p)})}{\pi b_0^2} \Theta(b_0 - b_j), \tag{2.13}$$

where n is the total number of trajectories, $\varepsilon_j^{(p)}$ is the relative kinetic energy of the proton at time t_j, b_j is the impact parameter at time t_j, and $\Theta(\cdot)$ is the step function. For a particular trajectory, t_j is defined as the instant in time when the proton-^{15}N distance takes its smallest value. Equation (2.13) implies that each trajectory with impact parameter smaller than b_0 contributes $(1/n)\sigma/(\pi b_0^2)$ to the total probability.

The result of our calculations is summarized in Fig. 2.5. In Fig. 2.5a, we show the probability for triggering the $p + {}^{15}\mathrm{N} \rightarrow {}^{12}\mathrm{C} + \alpha$ reaction for one molecule and one laser pulse, as a function of the CEP of the pulse. For each data point, 10^6 trajectories were calculated. As is known from studies on the electron recollision process [126, 127], the kinetic energy of the recolliding particle depends strongly on the CEP. In our case, also the number of recollision trajectories is highly CEP dependent, due to our particular setup of the proton starting to move at a particular distance (around 1 Å, which is the equilibrium internuclear distance in neutral NH) away from the center at which it later recollides. Moreover, due to this initial separation, it is crucial to include the contribution from the **B** field in the equations of motion. The effect of the $(\mathbf{v}/c) \times \mathbf{B}$ force, pushing the particles in the propagation direction $\hat{\mathbf{k}}$ of the laser pulse, can be clearly seen in Fig. 2.5b, where an example of a proton recollision trajectory is displayed. Both the proton and the $^{15}\mathrm{N}^{7+}$ nucleus are accelerated by the strong laser field, mostly along the laser polarization direction, but also slightly along the propagation direction, which in this case coincides with the orientation of the molecular axis. It is clear that without the slight push in the propagation direction, no recollision would take place. The larger drift in the propagation direction of the proton

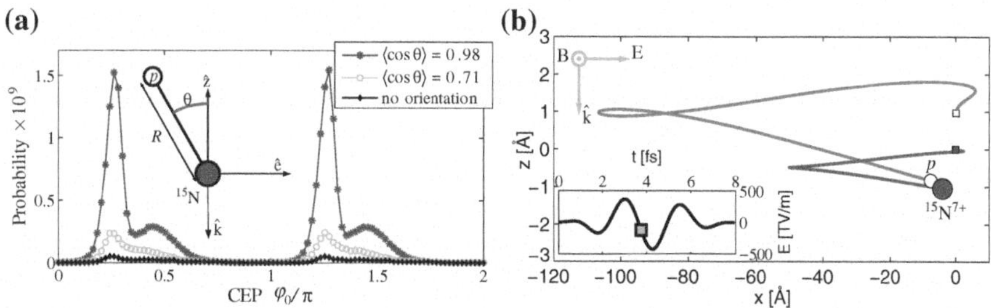

Fig. 2.5 **a** Total probability of the $p + {}^{15}\mathrm{N} \rightarrow {}^{12}\mathrm{C} + \alpha$ reaction, per molecule and per laser pulse, as a function of the CEP. Three different distributions for the orientation angle θ are considered. The *inset* defines the relevant angles. The propagation direction of the laser pulse is indicated by $\hat{\mathbf{k}}$. **b** Recollision trajectory. The initial positions of the proton and the ${}^{15}\mathrm{N}^{7+}$ nucleus are indicated by an *open square* and a *solid square*, respectively. The *inset* shows the laser field $E_0\Phi(t)$ (the CEP $\varphi_0 = 0.26\pi$), with the recollision instant indicated with a *filled square*. In both **a** and **b**, the laser field intensity is $I_0 = 2.5 \times 10^{22}$ W/cm^2, the wavelength is 800 nm, and the number of field cycles is $n_{\mathrm{cyc}} = 3$

compared to the ${}^{15}\mathrm{N}^{7+}$ nucleus also explains the benefit of aligning the molecular axis along the $\hat{\mathbf{k}}$ direction with the ${}^{15}\mathrm{N}$ atom pointing in the positive $\hat{\mathbf{k}}$ direction.

We conclude this section with the remark that the total probability for triggering the $p + {}^{15}\mathrm{N} \rightarrow {}^{12}\mathrm{C} + \alpha$ reaction is rather small, around 10^{-9} per molecule and per laser shot, as can be seen in Fig. 2.5a. In order to observe the proposed process experimentally, a laser system with high repetition rate or a dense molecular sample should be used in order to increase the number of events.

2.5 Summary

The classical trajectory model is a very useful one in the context of laser-atom and laser-molecule interaction. Large numerical grids which are required for a quantum mechanical simulation can be avoided. Once several trajectories are calculated, those which lead to a specific final outcome can be analyzed separately, often leading to an identification of the underlying physical mechanism. A good example of this procedure is the explanation of the high-energy peak of the $\mathrm{D}^+ + \mathrm{D}^+ + \mathrm{D}$ channel in the spectrum of the released kinetic energy of D_3^+ (see Sect. 2.3.2).

In this review, we have treated classical models which are *complete*, meaning that all interactions are included within the same theoretical model. An exception was the nuclear reaction described in Sect. 2.4.3, where an experimentally measured cross section was invoked to estimate the reaction probability. A complete model is able to describe all aspects of laser-atom or laser-molecule interaction: The stable ground state before the interaction with the external field, the time-dependent dynamics of the interacting many-body system including various types of particle rearrangement, as well as the final state after the interaction in which the system is in most cases decomposed into several smaller pieces. However, while the single-particle interaction with the external field can be described to a very good approximation in classical mechanics, this is not necessarily so for the interparticle interactions leading to particle rearrangement. Thus, for most cases one cannot hope to make quantitative comparisons with experimentally measured data. On the other hand, in the case of the highly complex dynamics of many-body systems interacting with laser light, we may seek a qualitative explanation, or a physical picture of the process at hand. In this respect, classical trajectory methods can be of great help, especially in the case when the system is so complex so that it is difficult to obtain even a qualitative picture with quantum mechanical methods.

Acknowledgments The research reported in Sects. 2.3.2 and 2.4.1 was supported by the Ministry of Education, Culture, Sports, Science and Technology (MEXT), Japan (Grant-in-Aid for Specially Promoted Research on Ultrafast Hydrogen Migration No.19002006), the Japan Society for the Promotion of Science (Grant-in-Aid for Challenging Exploratory Research No. 24655005 and Grant-in-Aid for Scientific Research No. 21-09238), and the Global COE Program "Chemistry Innovation through Cooperation of Science and Engineering" of the University of Tokyo. The numerical calculations reported in Sects. 2.3.1, 2.4.2, and 2.4.3 were carried out at the RIKEN Integrated Cluster of Clusters.

References

1. A. l'Huillier, L.A. Lompre, G. Mainfray, C. Manus, Phys. Rev. A **27**, 2503 (1983)
2. D.N. Fittinghoff, P.R. Bolton, B. Chang, K.C. Kulander, Phys. Rev. Lett. **69**, 2642 (1992)
3. K. Kondo, A. Sagisaka, T. Tamida, Y. Nabekawa, S. Watanabe, Phys. Rev. A **48**, R2531 (1993)
4. B. Walker, B. Sheehy, L.F. DiMauro, P. Agostini, K.J. Schafer, K.C. Kulander, Phys. Rev. Lett. **73**, 1227 (1994)
5. C. Figueira de Morisson Faria, X. Liu. J. Mod. Opt. **58**, 1076 (2011)
6. W. Becker, X. Liu, P.J. Ho, J.H. Eberly, Rev. Mod. Phys. **84**, 1011 (2012)
7. X. Sun, M. Li, D. Ye, G. Xin, L. Fu, X. Xie, Y. Deng, C. Wu, J. Liu, Q. Gong, Y. Liu, Phys. Rev. Lett. **113**, 103001 (2014)
8. M.Y. Kuchiev, JETP Lett. **45**, 404 (1987)
9. P.B. Corkum, Phys. Rev. Lett. **71**, 1994 (1993)
10. S. Augst, A. Talebpour, S.L. Chin, Y. Beaudoin, M. Chaker, Phys. Rev. A **52**, R917 (1995)
11. S. Larochelle, A. Talebpour, S.L. Chin, J. Phys. B **31**, 1201 (1998)
12. R. Moshammer, B. Feuerstein, W. Schmitt, A. Dorn, C.D. Schröter, J. Ullrich, H. Rottke, C. Trump, M. Wittmann, G. Korn, K. Hoffmann, W. Sandner, Phys. Rev. Lett. **84**, 447 (2000)
13. H. Maeda, M. Dammasch, U. Eichmann, W. Sandner, A. Becker, F.H.M. Faisal, Phys. Rev. A **62**, 035402 (2000)
14. A. Rudenko, K. Zrost, B. Feuerstein, V.L.B. de Jesus, C.D. Schröter, R. Moshammer, J. Ullrich, Phys. Rev. Lett. **93**, 253001 (2004)
15. K. Yamakawa, Y. Akahane, Y. Fukuda, M. Aoyama, N. Inoue, H. Ueda, T. Utsumi, Phys. Rev. Lett. **92**, 123001 (2004)
16. S. Palaniyappan, A. DiChiara, E. Chowdhury, A. Falkowski, G. Ongadi, E.L. Huskins, B.C. Walker, Phys. Rev. Lett. **94**, 243003 (2005)
17. K. Zrost, A. Rudenko, T. Ergler, B. Feuerstein, V.L.B. de Jesus, C.D. Schröter, R. Moshammer, J. Ullrich, J. Phys. B **39**, S371 (2006)

18. S. Palaniyappan, A. DiChiara, I. Ghebregziabher, E.L. Huskins, A. Falkowski, D. Pajerowski, B.C. Walker, J. Phys. B **39**, S357 (2006)
19. A. Rudenko, T. Ergler, K. Zrost, B. Feuerstein, V.L.B. de Jesus, C.D. Schröter, R. Moshammer, J. Ullrich, J. Phys. B **41**, 081006 (2008)
20. B.K. McFarland, J.P. Farrell, P.H. Bucksbaum, M. Gühr, Science **322**, 1232 (2008)
21. O. Smirnova, Y. Mairesse, S. Patchkovskii, N. Dudovich, D. Villeneuve, P. Corkum, M.Y. Ivanov, Nature (London) **460**, 972 (2009)
22. A.D. Shiner, B.E. Schmidt, C. Trallero-Herrero, H.J. Wörner, S. Patchkovskii, P.B. Corkum, J.-C. Kieffer, F. Légaré, D.M. Villeneuve, Nat. Phys. **7**, 464 (2011)
23. J. McKenna, A.M. Sayler, B. Gaire, N.G. Johnson, K.D. Carnes, B.D. Esry, I. Ben-Itzhak, Phys. Rev. Lett. **103**, 103004 (2009)
24. A.M. Sayler, J. McKenna, B. Gaire, N.G. Kling, K.D. Carnes, I. Ben-Itzhak, Phys. Rev. A **86**, 033425 (2012)
25. A.M. Sayler, J. McKenna, B. Gaire, N.G. Kling, K.D. Carnes, B.D. Esry, I. Ben-Itzhak, J. Phys. B **47**, 031001 (2014)
26. K. Hosaka, A. Yokoyama, K. Yamanouchi, R. Itakura, J. Chem. Phys. **138**, 204301 (2013)
27. C. Wang, B. Wang, M. Okunishi, W. Roeterdink, D. Ding, R. Zhu, G. Prümper, K. Shimada, K. Ueda, Chem. Phys. **430**, 40 (2014)
28. S. Roither, X. Xie, D. Kartashov, L. Zhang, M. Schöffler, H. Xu, A. Iwasaki, T. Okino, K. Yamanouchi, A. Baltuska, M. Kitzler, Phys. Rev. Lett. **106**, 163001 (2011)
29. X. Gong, Q. Song, Q. Ji, H. Pan, J. Ding, J. Wu, H. Zeng, Phys. Rev. Lett. **112**, 243001 (2014)
30. M.F. Kling, C. Siedschlag, A.J. Verhoef, J.I. Khan, M. Schultze, T. Uphues, Y. Ni, M. Uiberacker, M. Drescher, F. Krausz, M.J.J. Vrakking, Science **312**, 246 (2006)
31. H. Xu, J.-P. Maclean, D.E. Laban, W.C. Wallace, D. Kielpinski, R.T. Sang, I.V. Litvinyuk, New J. Phys. **15**, 023034 (2013)
32. I. Znakovskaya, P. von den Hoff, N. Schirmel, G. Urbasch, S. Zherebtsov, B. Bergues, R. de Vivie-Riedle, K.-M. Weitzel, M.F. Kling, Phys. Chem. Chem. Phys. **13**, 8653 (2011)
33. S. Miura, T. Ando, K. Ootaka, A. Iwasaki, H. Xu, T. Okino, K. Yamanouchi, D. Hoff, T. Rathje, G.G. Paulus, M. Kitzler, A. Baltuška, G. Sansone, M. Nisoli, Chem. Phys. Lett. **595–596**, 61 (2014)
34. A.S. Alnaser, M. Kübel, R. Siemering, B. Bergues, N.G. Kling, K.J. Betsch, Y. Deng, J. Schmidt, Z. Alahmed, A. Azzeer, J. Ullrich, I. Ben-Itzhak, R. Moshammer, U. Kleineberg, F. Krausz, R. de Vivie-Riedle, M. Kling, Nat. Commun. **5**, 3800 (2014)
35. X. Xie, K. Doblhoff-Dier, H. Xu, S. Roither, M.S. Schöffler, D. Kartashov, S. Erattupuzha, T. Rathje, G.G. Paulus, K. Yamanouchi, A. Baltuška, S. Gräfe, M. Kitzler, Phys. Rev. Lett. **112**, 163003 (2014)
36. X. Xie, S. Roither, M. Schöffler, E. Lötstedt, D. Kartashov, L. Zhang, G.G. Paulus, A. Iwasaki, A. Baltuška, K. Yamanouchi, M. Kitzler, Phys. Rev. X **4**, 021005 (2014)
37. R. Itakura, K. Yamanouchi, T. Tanabe, T. Okamoto, F. Kannari, J. Chem. Phys. **119**, 4179 (2003)
38. H. Kono, Y. Sato, N. Tanaka, T. Kato, K. Nakai, S. Koseki, Y. Fujimura, Chem. Phys. **304**, 203 (2004)
39. H. Yazawa, T. Tanabe, T. Okamoto, M. Yamanaka, F. Kannari, R. Itakura, K. Yamanouchi, J. Chem. Phys. **124**, 204314 (2006)
40. H. Yazawa, T. Shioyama, H. Hashimoto, F. Kannari, R. Itakura, K. Yamanouchi, Appl. Phys. B **98**, 275 (2010)
41. T. Ikuta, K. Hosaka, H. Akagi, A. Yokoyama, K. Yamanouchi, F. Kannari, R. Itakura, J. Phys. B **44**, 191002 (2011)
42. J.S. Parker, K.J. Meharg, G.A. McKenna, K.T. Taylor, J. Phys. B **40**, 1729 (2007)
43. S.X. Hu, Phys. Rev. Lett. **111**, 123003 (2013)
44. M. Lein, S. Kümmel, Phys. Rev. Lett. **94**, 143003 (2005)
45. F. Wilken, D. Bauer, Phys. Rev. Lett. **97**, 203001 (2006)
46. M. Thiele, E.K.U. Gross, S. Kümmel, Phys. Rev. Lett. **100**, 153004 (2008)
47. J. Heslar, D.A. Telnov, S.-I. Chu, Phys. Rev. A **87**, 052513 (2013)

48. T. Kato, H. Kono, Chem. Phys. Lett. **392**, 533 (2004)
49. T. Kato, H. Kono, J. Chem. Phys. **128**, 184102 (2008)
50. L. Greenman, P.J. Ho, S. Pabst, E. Kamarchik, D.A. Mazziotti, R. Santra, Phys. Rev. A **82**, 023406 (2010)
51. D. Hochstuhl, M. Bonitz, J. Chem. Phys. **134**, 084106 (2011)
52. D. Hochstuhl, M. Bonitz, Phys. Rev. A **86**, 053424 (2012)
53. H. Miyagi, L.B. Madsen, Phys. Rev. A **87**, 062511 (2013)
54. T. Sato, K.L. Ishikawa, Phys. Rev. A **88**, 023402 (2013)
55. H. Miyagi, L.B. Madsen, Phys. Rev. A **89**, 063416 (2014)
56. H. Miyagi, L.B. Madsen, J. Chem. Phys. **140**, 164309 (2014)
57. T. Sato, K.L. Ishikawa, Phys. Rev. A **91**, 023417 (2015)
58. M.A. Lysaght, H.W. van der Hart, P.G. Burke, Phys. Rev. A **79**, 053411 (2009)
59. M.A. Lysaght, S. Hutchinson, H.W. van der Hart, New J. Phys. **11**, 093014 (2009)
60. S. Hutchinson, M.A. Lysaght, H.W. van der Hart, Phys. Rev. A **88**, 023424 (2013)
61. T. Fennel, K.-H. Meiwes-Broer, J. Tiggesbäumker, P.-G. Reinhard, P.M. Dinh, E. Suraud, Rev. Mod. Phys. **82**, 1793 (2010)
62. R. Abrines, I.C. Percival, Proc. Phys. Soc. **88**, 861 (1966)
63. J.G. Leopold, I.C. Percival, Phys. Rev. Lett. **41**, 944 (1978)
64. H. Bauke, H.G. Hetzheim, G.R. Mocken, M. Ruf, C.H. Keitel, Phys. Rev. A **83**, 063414 (2011)
65. D. Bauer, Phys. Rev. A **56**, 3028 (1997)
66. R. Panfili, S.L. Haan, J.H. Eberly, Phys. Rev. Lett. **89**, 113001 (2002)
67. P.J. Ho, R. Panfili, S.L. Haan, J.H. Eberly, Phys. Rev. Lett. **94**, 093002 (2005)
68. P.J. Ho, J.H. Eberly, Phys. Rev. Lett. **95**, 193002 (2005)
69. S.L. Haan, J.S. Van Dyke, Z.S. Smith, Phys. Rev. Lett. **101**, 113001 (2008)
70. X. Wang, J.H. Eberly, Phys. Rev. Lett. **103**, 103007 (2009)
71. Y. Zhou, Q. Liao, P. Lu, Phys. Rev. A **80**, 023412 (2009)
72. F. Mauger, C. Chandre, T. Uzer, Phys. Rev. Lett. **102**, 173002 (2009)
73. F. Mauger, C. Chandre, T. Uzer, J. Phys. B **42**, 165602 (2009)
74. F. Mauger, C. Chandre, T. Uzer, Phys. Rev. Lett. **104**, 043005 (2010)
75. F. Mauger, C. Chandre, T. Uzer, Phys. Rev. A **81**, 063425 (2010)
76. X. Wang, J.H. Eberly, Phys. Rev. Lett. **105**, 083001 (2010)
77. A. Kamor, F. Mauger, C. Chandre, T. Uzer, Phys. Rev. E **83**, 036211 (2011)
78. X. Wang, J.H. Eberly, Phys. Rev. A **86**, 013421 (2012)
79. X. Wang, J.H. Eberly, J. Chem. Phys. **137**, 22A542 (2012)
80. X. Wang, J. Tian, J.H. Eberly, Phys. Rev. Lett. **110**, 073001 (2013)
81. E. Lötstedt, K. Midorikawa, Phys. Rev. A **87**, 013426 (2013)
82. Q. Tang, C. Huang, Y. Zhou, P. Lan, P. Lu, Phys. Rev. A **89**, 053419 (2014)
83. T. Wang, X.-L. Ge, J. Guo, X.-S. Liu, Phys. Rev. A **90**, 033420 (2014)
84. P.J. Ho, J.H. Eberly, Phys. Rev. Lett. **97**, 083001 (2006)
85. Y. Zhou, Q. Liao, P. Lu, Opt. Express **18**, 16025 (2010)
86. Q. Tang, C. Huang, Y. Zhou, P. Lu, Opt. Express **21**, 21433 (2013)
87. J.S. Cohen, Phys. Rev. A **64**, 043412 (2001)
88. T. Brabec, M.Y. Ivanov, P.B. Corkum, Phys. Rev. A **54**, R2551 (1996)
89. J. Chen, J. Liu, W.M. Zheng, Phys. Rev. A **66**, 043410 (2002)
90. D.F. Ye, X. Liu, J. Liu, Phys. Rev. Lett. **101**, 233003 (2008)
91. M.Y. Wu, Y.L. Wang, X.J. Liu, W.D. Li, X.L. Hao, J. Chen, Phys. Rev. A **87**, 013431 (2013)
92. Y. Liu, L. Fu, D. Ye, J. Liu, M. Li, C. Wu, Q. Gong, R. Moshammer, J. Ullrich, Phys. Rev. Lett. **112**, 013003 (2014)
93. M.V. Ammosov, N.B. Delone, V.P. Krainov, Sov. Phys. JETP **64**, 1191 (1986)
94. L. Wilets, E. Henley, M. Kraft, A. Mackellar, Nucl. Phys. A **282**, 341 (1977)
95. C.L. Kirschbaum, L. Wilets, Phys. Rev. A **21**, 834 (1980)
96. W.H. Press, S.A. Teukolsky, W.T. Vetterling, B.P. Flannery, *Numerical Recipes in Fortran 77: The Art of Scientific Computing* (Cambridge University Press, Cambridge, 1992)

97. J.S. Cohen, Phys. Rev. A **56**, 3583 (1997)
98. J.S. Cohen, Phys. Rev. A **51**, 266 (1995)
99. J.S. Cohen, Phys. Rev. A **57**, 4964 (1998)
100. J.A. Nelder, R. Mead, Comput. J. **7**, 308 (1965)
101. D. Zajfman, D. Maor, Phys. Rev. Lett. **56**, 320 (1986)
102. W.A. Beck, L. Wilets, M.A. Alberg, Phys. Rev. A **48**, 2779 (1993)
103. W.A. Beck, L. Wilets, Phys. Rev. A **55**, 2821 (1997)
104. J.S. Cohen, Phys. Rev. A **59**, 4300 (1999)
105. J.S. Cohen, Phys. Rev. A **59**, 1160 (1999)
106. J.S. Cohen, J. Phys. B **38**, 441 (2005)
107. J.S. Cohen, J. Phys. B **39**, 3561 (2006)
108. J.S. Cohen, J. Phys. B **39**, 1517 (2006)
109. D.A. Wasson, S.E. Koonin, Phys. Rev. A **39**, 5676 (1989)
110. P.B. Lerner, K.J. LaGattuta, J.S. Cohen, Laser Phys. **3**, 331 (1993)
111. K.J. LaGattuta, J.S. Cohen, J. Phys. B **31**, 5281 (1998)
112. K.J. LaGattuta, J. Phys. B **33**, 2489 (2000)
113. Y. Zhou, C. Huang, Q. Liao, P. Lu, Phys. Rev. Lett. **109**, 053004 (2012)
114. Y. Zhou, C. Huang, P. Lu, Opt. Express **20**, 20201 (2012)
115. Y. Zhou, Q. Zhang, C. Huang, P. Lu, Phys. Rev. A **86**, 043427 (2012)
116. K.J. LaGattuta, Phys. Rev. A **73**, 043404 (2006)
117. E. Lötstedt, T. Kato, K. Yamanouchi, Phys. Rev. Lett. **106**, 203001 (2011)
118. C. Huang, Z. Li, Y. Zhou, Q. Tang, Q. Liao, P. Lu, Opt. Express **20**, 11700 (2012)
119. E. Lötstedt, T. Kato, K. Yamanouchi, Phys. Rev. A **85**, 053410 (2012)
120. D. Bauer, J. Phys. B **37**, 3085 (2004)
121. L. Wilets, J.S. Cohen, Contemp. Phys. **39**, 163 (1998)
122. K.J. LaGattuta, J. Phys. A **36**, 6013 (2003)
123. E. Lötstedt, K. Midorikawa, Phys. Rev. A **90**, 043415 (2014)
124. A. Becker, F.H.M. Faisal, Y. Liang, S. Augst, Y. Beaudoin, M. Chaker, S.L. Chin, J. Phys. B **33**, L547 (2000)
125. G. Marcus, W. Helml, X. Gu, Y. Deng, R. Hartmann, T. Kobayashi, L. Strueder, R. Kienberger, F. Krausz, Phys. Rev. Lett. **108**, 023201 (2012)
126. X. Liu, C. Figueira de Morisson, Faria. Phys. Rev. Lett. **92**, 133006 (2004)
127. B. Bergues, M. Kübel, N.G. Johnson, B. Fischer, N. Camus, K.J. Betsch, O. Herrwerth, A. Senftleben, A.M. Sayler, T. Rathje, T. Pfeifer, I. Ben-Itzhak, R.R. Jones, G.G. Paulus, F. Krausz, R. Moshammer, J. Ullrich, M.F. Kling, Nat. Commun. **3**, 813 (2012)
128. J.D. Alexander, C.R. Calvert, R.B. King, O. Kelly, L. Graham, W.A. Bryan, G.R.A.J. Nemeth, W.R. Newell, C.A. Froud, I.C.E. Turcu, E. Springate, I.D. Williams, J.B. Greenwood, J. Phys. B **42**, 141004 (2009)
129. J. McKenna, A.M. Sayler, B. Gaire, N.G. Kling, B.D. Esry, K.D. Carnes, I. Ben-Itzhak, New J. Phys. **14**, 103029 (2012)
130. B. Gaire, J. McKenna, M. Zohrabi, K.D. Carnes, B.D. Esry, I. Ben-Itzhak, Phys. Rev. A **85**, 023419 (2012)
131. H. Yu, A.D. Bandrauk, J. Chem. Phys. **102**, 1257 (1995)
132. H. Yu, A.D. Bandrauk, Phys. Rev. A **56**, 685 (1997)
133. A.D. Bandrauk, H. Yu, Phys. Rev. A **59**, 539 (1999)
134. I. Kawata, H. Kono, A.D. Bandrauk, Phys. Rev. A **64**, 043411 (2001)
135. D.S. Tchitchekova, H. Lu, S. Chelkowski, A.D. Bandrauk, J. Phys. B **44**, 065601 (2011)
136. M. Isla, J.A. Alonso, J. Phys. Chem. C **111**, 17765 (2007)
137. D. Frye, A. Preiskorn, G.C. Lie, E. Clementi, J. Chem. Phys. **92**, 4948 (1990)
138. B. Manschwetus, T. Nubbemeyer, K. Gorling, G. Steinmeyer, U. Eichmann, H. Rottke, W. Sandner, Phys. Rev. Lett. **102**, 113002 (2009)
139. E. Lötstedt, T. Kato, K. Yamanouchi, J. Phys. B **46**, 235601 (2013)
140. D.M. Villeneuve, M.Y. Ivanov, P.B. Corkum, Phys. Rev. A **54**, 736 (1996)
141. W. Qu, S. Hu, Z. Xu, Phys. Rev. A **57**, 4528 (1998)

142. F. Mauger, C. Chandre, T. Uzer, Chem. Phys. **366**, 64 (2009)
143. A. Di Piazza, C. Müller, K.Z. Hatsagortsyan, C.H. Keitel, Rev. Mod. Phys. **84**, 1177 (2012)
144. M. Verschl, C.H. Keitel, J. Phys. B **40**, F69 (2007)
145. N. Milosevic, P.B. Corkum, T. Brabec, Phys. Rev. Lett. **92**, 013002 (2004)
146. J.D. Jackson, *Classical Electrodynamics* (Wiley, Hoboken, 1998)
147. A.N. Kaufman, P.S. Rostler, Phys. Fluids **14**, 446 (1971)
148. T. Popmintchev, M.-C. Chen, D. Popmintchev, P. Arpin, S. Brown, S. Ališauskas, G. Andriukaitis, T. Balčiunas, O.D. Mücke, A. Pugzlys, A. Baltuška, B. Shim, S.E. Schrauth, A. Gaeta, C. Hernández-García, L. Plaja, A. Becker, A. Jaron-Becker, M.M. Murnane, H.C. Kapteyn, Science **336**, 1287 (2012)
149. C. Hernández-García, J.A. Pérez-Hernández, T. Popmintchev, M.M. Murnane, H.C. Kapteyn, A. Jaron-Becker, A. Becker, L. Plaja, Phys. Rev. Lett. **111**, 033002 (2013)
150. M.-C. Chen, C. Mancuso, C. Hernández-García, F. Dollar, B. Galloway, D. Popmintchev, P.-C. Huang, B. Walker, L. Plaja, A.A. Jaroń-Becker, A. Becker, M.M. Murnane, H.C. Kapteyn, T. Popmintchev, Proc. Natl. Acad. Sci. USA **111**, E2361 (2014)
151. A.-T. Le, H. Wei, C. Jin, V.N. Tuoc, T. Morishita, C.D. Lin, Phys. Rev. Lett. **113**, 033001 (2014)
152. M. Klaiber, K.Z. Hatsagortsyan, C.H. Keitel, Phys. Rev. A **75**, 063413 (2007)
153. H.R. Reiss, Phys. Rev. A **82**, 023418 (2010)
154. E. Lötstedt, K. Midorikawa, Phys. Rev. Lett. **112**, 093001 (2014)
155. O. Smirnova, M. Spanner, M. Ivanov, Phys. Rev. Lett. **90**, 243001 (2003)
156. H. Niikura, F. Légaré, R. Hasbani, M.Y. Ivanov, D.M. Villeneuve, P.B. Corkum, Nature (London) **421**, 826 (2003)
157. S. Chelkowski, A.D. Bandrauk, P.B. Corkum, Phys. Rev. Lett. **93**, 083602 (2004)
158. A.V. Sokolov, M. Zhi, J. Mod. Opt. **51**, 2607 (2004)
159. G.K. Paramonov, Chem. Phys. **338**, 329 (2007)
160. M. Zhi, A.V. Sokolov, Phys. Rev. A **80**, 023415 (2009)
161. H.M. Castañeda Cortés, C. Müller, C.H. Keitel, A. Pálffy, Phys. Lett. B **723**, 401 (2013)
162. V. Yanovsky, V. Chvykov, G. Kalinchenko, P. Rousseau, T. Planchon, T. Matsuoka, A. Maksimchuk, J. Nees, G. Cheriaux, G. Mourou, K. Krushelnick, Opt. Express **16**, 2109 (2008)
163. A. Redder, H.W. Becker, H. Lorenz-Wirzba, C. Rolfs, P. Schmalbrock, H.P. Trautvetter, Z. Phys. A **305**, 325 (1982)
164. O. Ghafur, A. Rouzée, A. Gijsbertsen, W. Kiu Siu, S. Stolte, M.J.J. Vrakking Nat. Phys. **5**, 289 (2009)
165. K. Oda, M. Hita, S. Minemoto, H. Sakai, Phys. Rev. Lett. **104**, 213901 (2010)

Chapter 3
Nonadiabatic Molecular Alignment and Orientation

Hirokazu Hasegawa and Yasuhiro Ohshima

Abstract In this chapter, we review recent researches on nonadiabatic molecular alignment and orientation induced by intense nonresonant short laser pulses. In addition to a typical femtosecond pump-probe method to probe dynamics of a rotational wave packet, a novel method to study molecular alignment is introduced. We focus on recent spectroscopic studies carried out in the frequency domain that demonstrate that both amplitude (population) and phase of the rotational wave packet can be reconstructed by combining femtosecond and nanosecond pulses.We also demonstrate the creation of the peculiar rotational wave packet corresponding to right- or left-handed rotation.

3.1 Introduction

When an intense laser pulse with linear polarization is applied to molecules in gas phase, molecules receive strong torque generated by the interaction of the laser field with an induced dipole moment even if the laser's wavelength is nonresonant. The torque forces to align parallel to the laser polarization direction. If the pulse duration of the incident pulse is shorter than the period of molecular rotational motion, typically, several ten picoseconds, molecules keep rotating after the interaction. Consequently, a molecular axis distribution changes as a function of time. During this dynamics, the molecular axis distribution is spatially localized at specific times periodically. This phenomenon is called nonadiabatic molecular alignment (NAMA) [1–3].

H. Hasegawa (✉)
The University of Tokyo, 3-8-1 Komaba, Meguro-ku, Tokyo 153-8902, Japan
e-mail: chs36@mail.ecc.u-tokyo.ac.jp

Y. Ohshima
Tokyo Institute of Technology, 2-12-1 Ookayama, Meguro-ku, Tokyo 152-8550, Japan
e-mail: ohshima@chem.titech.ac.jp

K. Yamanouchi et al. (eds.), *Progress in Ultrafast Intense Laser Science XII*,
Springer Series in Chemical Physics 112, DOI 10.1007/978-3-319-23657-5_3

The NAMA has been experimentally studied during last decades by using methods such as Coulomb explosion [4–6], transient birefringence [7–10], high-order harmonics [11–15], four-wave mixing [16–19], and population measurement [20–26] in addition to theoretical researches [27–36]. The majority of molecular species is linear molecule with a center of symmetry like N_2, CO_2, and C_2H_2. There are relatively few studies for symmetric [22–24, 26, 37, 38] and asymmetric molecules [39–47] because it is hard how to implement it and evaluate the extent of the NAMA [46]. In addition to these rigid molecules, the NAMA of weakly bounded clusters has been reported [48–50].

From the standpoint of applications utilizing the NAMA, the higher extent of alignment is desirable. It is, however, difficult to increase the extent of the molecular alignment by increasing a laser intensity due to ionization. The efforts to enhance and control the extent of the alignment have also been performed by using multi-pulse [33, 38, 51–55] and shaped pulse [36, 56–60].

The NAMA is also available for applications such as isotope separation [17, 18, 61, 62], ionization probability [63–69], molecular orbital tomography [11, 70], molecular deformation [71], selective control of fragmentation reactions [72], and electron diffraction [73].

In the case of linear molecules without a center of symmetry like CO, it is important to discriminate the head-to-tail order of the molecule, which is called the nonadiabatic molecular orientation (NAMO). The NAMO has been also demonstrated by using methods like a short THz radiation [74–77], a combination of static field [78] and two-color fields [79, 80]. Since the property of molecule depends strongly on molecular orientation in space, the control of molecular rotation using NAMA and NAMO plays important roles in stereochemical reactions [81] and the measurement of physical quantities of molecule in a molecular fixed frame [82].

3.2 Non-adiabatic Molecular Alignment and Orientation

The NAMA and NAMO have been already reviewed by many authors [1–3]. We briefly overview their theoretical aspect in this section.The NAMA and NAMO are regarded as the dynamics of the superposition state of rotational eigen states, that is, a rotational wave packet. We focus on excitation process, wave packet motion, and how to evaluate the degree of the NAMA and NAMO.

3.2.1 Interaction Potential

The interaction of molecules with a nonresonant short laser field can be expressed as follows [83].

$$\hat{V}(t) = -\boldsymbol{\mu} \cdot \boldsymbol{E}(t) - \frac{1}{2}\boldsymbol{\alpha}\boldsymbol{E}(t)\boldsymbol{E}(t) - \frac{1}{6}\boldsymbol{\beta}\boldsymbol{E}(t)\boldsymbol{E}(t)\boldsymbol{E}(t) + \cdots$$
$$\equiv \hat{V}_{\text{dip}}(t) + \hat{V}_{\text{pol}}(t) + \hat{V}_{\text{hyp}}(t) + \cdots, \tag{3.1}$$

where $\boldsymbol{\mu}$ is a permanent electric dipole moment, $\boldsymbol{\alpha}$ is a polarizability tensor, $\boldsymbol{\beta}$ is a hyperpolarizability tensor, and $\boldsymbol{E}(t)$ is an electric field. The electric field with a linearly polarized light along a Z-axis in a laboratory-fixed frame is only focused hereafter although a circular or elliptical polarized pulse can induce molecular alignment [84, 85]. In the case of a nonresonant laser pulse, the electric field is averaged over a fast carrier cycle because the carrier electric fields don't affect a rotational motion. If the electric field is assumed to be a gaussian pulse shape with a single center frequency, ω, it can be described by the following equation.

$$E_Z(t) = \varepsilon_\omega(t)\cos(\omega t + \phi_\omega) \tag{3.2}$$

where $E_Z(t)$ is a component along a Z-axis of the electric field, $\varepsilon_\omega(t)$ is an envelope function, and ϕ_ω is a carrier envelope phase. $\hat{V}_{\text{dip}}(t)$ and $\hat{V}_{\text{hyp}}(t)$ vanish after averaging over the fast carrier vibration. As a result, the interaction potential is represented like

$$\hat{V}(t) = -\frac{1}{4}\alpha_{ZZ}\varepsilon_\omega(t)^2, \tag{3.3}$$

where α_{ZZ} is a tensor component of α along the laboratory-fixed Z-axis. This interaction causes nonadiabatic molecular alignment.

If the electric field is assumed to have two-color center frequencies of ω and 2ω, i.e., if the electric field is written by

$$E_Z(t) = \varepsilon_\omega(t)\cos(\omega t + \phi_\omega) + \varepsilon_{2\omega}(t)\cos(2\omega t + \phi_{2\omega}), \tag{3.4}$$

$\hat{V}_{\text{hyp}}(t)$ doesn't wash out although $\hat{V}_{\text{dip}}(t)$ vanishes after temporal average. Consequently, the interaction is expressed by the following equation [86]

$$\hat{V}(t) = -\frac{1}{4}\alpha_{ZZ}\left\{\varepsilon_\omega(t)^2 + \varepsilon_{2\omega}(t)^2\right\} - \frac{1}{8}\beta_{ZZZ}\varepsilon_\omega(t)^2\varepsilon_{2\omega}(t)\cos\phi, \tag{3.5}$$

where β_{ZZZ} is a component along the laboratory-fixed Z-axis of the hyper polarizability and $\phi = \phi_{2\omega} - \phi_\omega$ is a relative phase. In this equation, the first term originated from $\hat{V}_{\text{pol}}$ and the second term derived from $\hat{V}_{\text{hyp}}$ cause nonadiabatic molecular alignment and orientation, respectively. For linear molecules, α_{ZZ} and β_{ZZZ} need to be rewritten to the component in a molecular-fixed frame as follows

$$\alpha_{ZZ} = \Delta\alpha\cos^2\theta + \alpha_\perp, \tag{3.6}$$
$$\beta_{ZZZ} = (\beta_\parallel - 3\beta_\perp)\cos^3\theta + 3\beta_\perp\cos\theta, \tag{3.7}$$

where $\alpha_\perp$ and $\alpha_\parallel$ are the polarizability components perpendicular and parallel to a molecular axis, respectively, $\Delta\alpha = \alpha_\parallel - \alpha_\perp$, $\beta_\perp$ and $\beta_\parallel$ are the hyperpolarizability components perpendicular and parallel to the molecular axis, respectively, and θ is the angle between the polarization direction and the molecular axis. The $V_{\text{dip}}(t)$ and $V_{\text{hyp}}(t)$ originate from the anisotropy of the molecular polarizability and hyperpolarizability, respectively. All molecular species except for molecules with spherical symmetry, therefore, can be aligned and oriented some extent by intense laser fields.

3.2.2 Rotational Wave Packet

The Schödinger equation of the rotational degree of freedom for linear molecules with no electric angular and spin momenta is given by

$$\hat{H}_{\text{rot}}|J, M\rangle = E_J|J, M\rangle, \tag{3.8}$$

where $\hat{H}_{\text{rot}}$ means a rotational Hamiltonian, $E_J = hcBJ(J+1)$ is a rotational eigen energy, $|J, M\rangle$ is an eigen state, $J = 0, 1, 2, \ldots$ is a quantum number of a rotational angular momentum, $M = -J, -J+1, \ldots, J-1, J$ is a projection of J onto the space-fixed Z-axis, B is a rotational constant in units of cm^{-1}, h and c are Planck constant and the speed of light in vacuum, respectively. Each eigen state has a definite parity of $(-1)^J$. The effects of centrifugal distortion are neglected because those of typical rigid linear molecules such as N_2, CO_2, and C_2H_2 are three-order smaller than $hcBJ(J+1)$ for low rotational states.

The interaction represented by (3.3) and (3.5) changes an initial rotational eigen state to a rotational wave packet, $|\psi(t)\rangle$, which is a superposition state of rotational eigen states. The creation and dynamics of the rotational wave packet is governed by the time-dependent Schödinger equation (TDSE)

$$i\hbar\frac{\partial}{\partial t}|\psi(t)\rangle = \hat{H}|\psi(t)\rangle, \tag{3.9}$$

where $\hat{H} = \hat{H}_{\text{rot}} + \hat{V}(t)$ is a total Hamiltonian. The rotational wave packet is expanded by the rotational eigen states as follows,

$$|\psi(t)\rangle = \sum_J c_J(t)e^{-i\omega_J t}|J, M\rangle, \tag{3.10}$$

where c_J are complex expansion coefficients, $\omega_J = E_J/\hbar$. Here, the initial value of M is conserved due to an axial symmetry of linear polarization. From (3.9) and (3.10), $c_J(t)$ follows the differential equation

$$\frac{d}{dt}c_J(t) = \frac{1}{i\hbar}\sum_{J'} c_{J'}(t)e^{-i\omega_{J'J}t}\langle JM|\hat{V}(t)|J'M\rangle. \tag{3.11}$$

By solving the above equation numerically, the time-dependent expansion coefficients, $c(t)$, are obtained. After the interaction with an intense laser pulse, the $c_J(t)$ become constants. The matrix elements of α_{ZZ} and β_{ZZZ} included in the $\langle JM|\hat{V}(t)|J'M\rangle$ are expressed by [87]

$$\langle J,M|\alpha_{ZZ}|J,M\rangle = \Delta\alpha\left\{\frac{2}{3}\frac{J(J+1)-3M^2}{(2J+3)(2J-1)}+\frac{1}{3}\right\}+\alpha_{\perp}, \tag{3.12}$$

$$\langle J+2,M|\alpha_{ZZ}|J,M\rangle = \frac{\Delta\alpha}{2J+3}\sqrt{\frac{\{(J+2)^2-M^2\}\{(J+1)^2-M^2\}}{(2J+5)(2J+1)}}, \tag{3.13}$$

$$\langle J+1,M|\beta_{ZZZ}|J,M\rangle = \frac{3}{(2J+5)(2J-1)}\sqrt{\frac{(J+1)^2-M^2}{(2J+3)(2J+1)}} \times\left\{\beta_{\parallel}(J^2+2J-M^2-1)+\beta_{\perp}(J^2+2J+3M^2-2)\right\}, \tag{3.14}$$

$$\langle J+3,M|\beta_{ZZZ}|J,M\rangle = \frac{(\beta_{\parallel}-3\beta_{\perp})}{(2J+5)(2J+3)} \times\sqrt{\frac{\{(J+3)^2-M^2\}\{(J+2)^2-M^2\}\{(J+1)^2-M^2\}}{(2J+7)(2J+1)}}. \tag{3.15}$$

The $V_{\text{dip}}(t)$ having a positive parity connects between states with the same parity. In other words, the parity selection rule originated from $V_{\text{dip}}(t)$ is $\pm \leftrightarrow \pm$ and $\pm \not\leftrightarrow \mp$. On the other hand, since the $V_{\text{hyp}}(t)$ has a negative parity, it connects between states with the different parity. The parity selection rule of $V_{\text{hyp}}(t)$ is $\pm \leftrightarrow \mp$ and $\pm \not\leftrightarrow \pm$. The NAMO can be induced by the $V_{\text{hyp}}(t)$ because the descrimination of head-to-tail direction is achieved by mixing the states with different parity. As a result, the implementation of the NAMO requires a two-color laser pulse or a method to connect between states with different parity.

The selection rules of the NAMA induced by $V_{\text{dip}}(t)$ are the same $\Delta J = \pm 2$ as rotational Raman excitation. The nature of the strong laser field, however, allows multiple excitation called nonadiabatic rotational excitation (NAREX) [20, 22]. Consequently, successive rotational excitation excites the initial lower rotational state to higher rotational states. The selection rules of the NAMO induced by $V_{\text{hyp}}(t)$ are $\Delta J = \pm 1, \pm 3$. In this case, higher rotational states are also created with the aid of multiple excitation.

It should be noted that the rotational wave packet has periodicity. The wave packet satisfies with $|\psi(t+T_{\text{rot}}\rangle = |\psi(t)\rangle$ when $T_{\text{rot}} = \frac{1}{2Bc}$ because of the regularity of the rotational eigen energy, $E_J = hcBJ(J+1)$. The T_{rot} is called a rotational period.

3.2.3 Evaluation of the NAMA and NAMO

The expectation value of $\cos^2\theta$, $\langle\cos^2\theta\rangle(t) = \langle\psi(t)|\cos^2\theta|\psi(t)\rangle$, is usually used as the evaluation of the degree of the NAMA, where θ is the angle between a polarization direction and molecular axis for linear molecule. The value of $\langle\cos^2\theta\rangle(t)$ is 1 for perfect alignment, 1/3 for isotropic, and 0 for perfect anti-alignment. In order to compare experimental data with calculated data, the initial thermal ensemble needs to be considered.

For example, Fig. 3.1a shows a calculated $\langle\cos^2\theta\rangle(t)$ of N_2 with the rotational temperature of 300 K. An intense laser field of 30 TW/cm^2 and 100 fs is irradiated at $t = 0$. The rotational period of N_2 is calculated to be $T_{\rm rot} = 8.38$ ps from the rotational constant of $B = 1.9896$ cm^{-1} in the electronic ground state [88]. Clear five peaks appear at $t = 0, 2.1, 4.2, 6.3$, and 8.4 ps with an interval of $\frac{T_{\rm rot}}{4} = 2.1$ ps. The second to fifth peaks are called a quarter revival, a half revival, a three-quarter revival, and a full revival, respectively.

Panels 3.1b–d show the calculated $\langle\cos^2\theta\rangle(t)$ in which the single initial state, $|J_i, M_i\rangle = |0,0\rangle, |1,0\rangle, |1,\pm1\rangle, |2,0\rangle, |2,\pm1\rangle$, and $|2,\pm2\rangle$, are used. Although these signals behave complexly, all signals are in-phase at the half and full revivals. Inversely, signals of even J_i and odd J_i are out-of-phase by π at the quarter and three-quarter revivals. In the case of N_2, since N nuclei has a nuclear spin of $I = 1$, a nuclear spin statistical weight of even J states is 2 times larger than that of odd

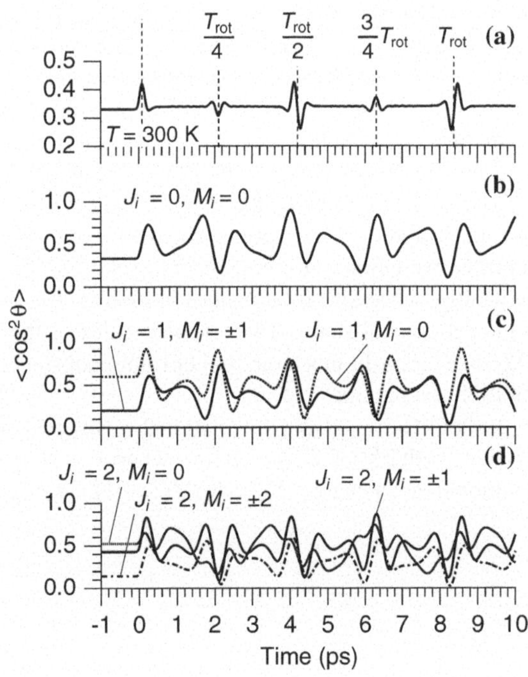

Fig. 3.1 The calculated $\langle\cos^2\theta\rangle(t)$ of N_2 irradiated by 30 TW/cm^2 and 100 fs. The initial state is **a** an ensemble of rotational temperature at 300 K, **b** $J = 0, M = 0$, **c** $J = 1$ with $M = 0, \pm1$, and **d** $J = 2$ with $M = 0, \pm1, \pm2$

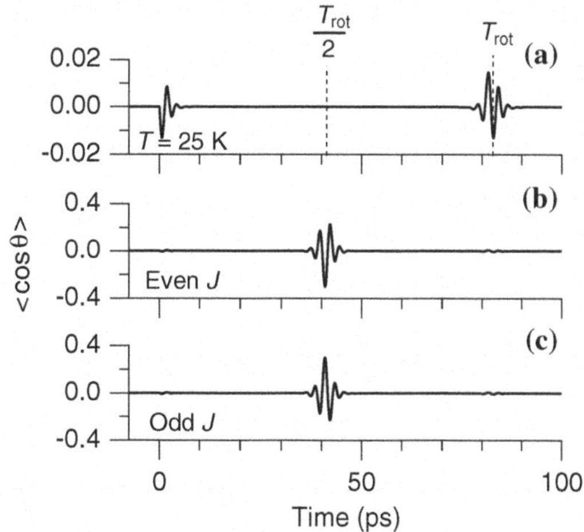

Fig. 3.2 The calculated $\langle\cos\theta\rangle(t)$ of OCS irradiated by 20 TW/cm^2 and 100 fs for both 800 and 400 nm. The initial state is **a** an ensemble of rotational temperature at 25 K, **b** only even J state in (**a**), and **c** only odd J state in (**a**)

J states. After averaging $\langle\cos^2\theta\rangle(t)$ over thermal distribution including the nuclear spin statistics, the signals at the half and full revivals enhances and those at the quarter and three-quarter revivals diminish as shown Fig. 3.1a.

The expectation value of $\cos\theta$, $\langle\cos\theta\rangle(t) = \langle\psi(t)|\cos\theta|\psi(t)\rangle$, is utilized for the evaluation of the degree of the NAMO. The value of $\langle\cos\theta\rangle(t)$ is 1 for perfect orientation (head-to-tail order), 0 for no orientation, and -1 for perfect orientation (tail-to-head order).

Figure 3.2a shows the calculated $\langle\cos\theta\rangle(t)$ of OCS with rotational temperature of 25 K irradiated by a two-color pulse, which is composed of a pulse with a duration of 100 fs, a center wavelength of 800 nm, and an intensity of 20 TW/cm^2 and a pulse with a duration of 100 fs, a center wavelength of 400 nm, and an intensity of 20 TW/cm^2. The relative phase of the two pulses is set to be $\phi = 0$. The rotational period of OCS, $T_{rot} = 82$ ps, is obtained from the rotational constant of $B = 0.202857$ cm^{-1} [89]. It is recognized that the two-color pulse induces the NAMO at the full revival time, $t = 82$ ps. Figure 3.2b and c show the $\langle\cos\theta\rangle(t)$ calculated from the initial states of even J and odd J, respectively. It should be noted that the scale of the vertical axes in Fig. 3.2b and c is 20 times larger than that in Fig. 3.2a. Although the large $\langle\cos\theta\rangle(t)$ value is achieved by the initial states of even or odd J, the degree of the orientation results in a smaller value because the summation of the contribution of the even J and odd J initial states, that is, thermal averaging.

3.3 Experimental Techniques to Pursue Rotational Wave Packet Dynamics

The dynamics of rotational wave packets has been experimentally investigated by using various methods as mentioned before. Coulomb explosion imaging, high-order harmonic generation, optical birefringence, and four-wave mixing are commonly used real-time measurements. In these method, a femtosecond pump pulse creates a rotational wave packet, and then another femtosecond probe pulse probes a molecular axis distribution by monitoring a angular distribution of a Coulomb exploded fragment ions or a polarization change or a generated radiation as a function of the delay between the pump and the probe pulse. In addition to these time domain experiments, a spectroscopic method has been reported in order to study the rotational wave packet motion [20–26, 68, 90].

3.3.1 Real Time Measurement

The first observation of the NAMA has been reported almost 40 years ago [91, 92]. Here, we introduce a recent study performed by Rosca-Pruna and Vrakking [4, 5, 93]. The NAMA of I_2 was induced by a intense laser pulse of 10^{13} W/cm^2 and 2.8 ps. The Coulomb explosion processes, $I_2^{2+} \rightarrow I^{2+} + I$, $I_2^{3+} \rightarrow I^{2+} + I^+$, and $I_2^{4+} \rightarrow I^{2+} + I^{2+}$, are induced by a delayed intense laser pulse of 5×10^{14} W/cm^2 and 100 fs. The molecular axis distributions were measured by observing angular distributions of the fragment ions with a two-dimensional ion detector. The dynamics of a rotational wave packet was studied on the basis of the angular distributions of the fragment ions as a function of the pump-probe delay.

The time-dependent angular distributions behaved like Fig. 3.1a with several sharp peaks. The observed intervals between neighboring peaks are about 110 ps. It is expected that a quarter- and a three-quarter-revival as well as a half- and full-revival appear because the nuclear spin weight of even J and odd J rotational states is 5:7 due to the nuclear spin of ^{127}I nuclei of $I = \frac{5}{2}$. Therefore, the intervals between neighboring peaks equal to $\frac{T_{\rm rot}}{4}$. The calculated value, $\frac{T_{\rm rot}}{4} = 112$ ps, from the rotational constant of $B = 0.03731$ cm^{-1} in the electronic ground state of I_2 [88] agrees with the observed interval. The fact gives clear evidence that the NAMA occurred.

3.3.2 Rotational Population of the Wave Packet in Frequency Domain

Molecular alignment of NO has been studied on the basis of frequency domain measurement [20]. In this experiment, ns laser pulses (~226 nm, an energy resolution of 0.4 cm^{-1}, 20 μJ/pulse, 10 ns duration) are used in order to probe the rotational population of NO molecules after the interaction with an intense laser field for an alignment pulse. Figure 3.3 shows excitation spectra recorded by (1 + 1) resonant-enhanced multiphoton ionization (REMPI) via the electronic $A^2\Sigma^+$ state measured by scanning the probe ns laser wavelength. The spectrum measured without pump pulses is shown in Fig. 3.3a. Since three peaks in Fig. 3.3a are assigned as $Q_{11}(0.5)$,

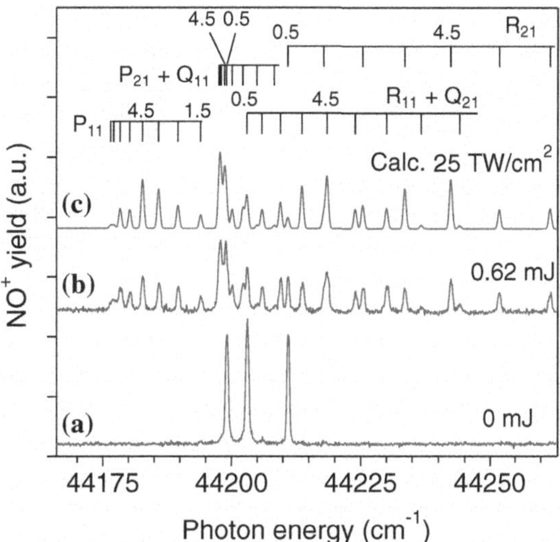

Fig. 3.3 The REMPI spectra of NO measured **a** without the pump pulse and **b** with the pump pulse. **c** The simulated spectrum

$R_{11}(0.5) + Q_{21}(0.5)$ and $R_{21}(0.5)$ transitions, the initial rotational state is concentrated on the lowest $J = 0.5$ state and then the rotational temperature is estimated to be lower than 2 K. The total angular momentum J has a half integer due to an unpaired electron in the $X^2\Pi_{1/2}$ electronic ground state.

Figure 3.3b shows the REMPI excitation spectrum measured with intense femtosecond laser pulses (0.62 mJ/pulse, 810 nm, 150 fs) irradiated 100 ns before the probe pulse. The peak around 44,244 cm^{-1} is assigned as R_{11}+ Q_{21}-branch with an initial state $J = 8.5$. The excitation of the rotational state up to $J = 8.5$ is clear evidence that the pump pulse excites rotational states.

In order to clarify that the rotational excitation is originated from NAREX, an excitation spectrum is simulated. The rotational wave packet generated from the initial state of $J = 0.5$ by the interaction with a laser field of 25 TW/cm^2 is calculated by numerically solving TDSE of (3.11). Then, the spectrum is obtained on the basis of both the transition probabilities of the A-X transition and the rotational state distribution of the wave packet. Figure 3.1c shows the simulated spectrum. The agreement between the observed spectrum and the calculated one supports that the observed rotational excitation results from NAREX.

The information on the rotational population is extracted from the observed peak intensities divided by the transition probabilities for each rotational line in the A-X transition. Figure 3.4 shows the rotational population obtained from the spectrum in Fig. 3.1b. The rotational distribution, in which states with $J = 2.5, 4.5$, and 6.5 have relatively a larger population than the adjacent $J = 1.5$, 3.5, and 5.5 states, is inherent in the excitation process and is explained as follows. The initial rotational state $J = 0.5$ is coupled with both $J = 1.5$ and $J = 2.5$ due to the selection rules of $\Delta J = \pm 1, \pm 2$ for NO molecules having an unpaired electron. The population of $J = 1.5$ and 2.5 transferred from $J = 0.5$ is excited to higher rotational states through

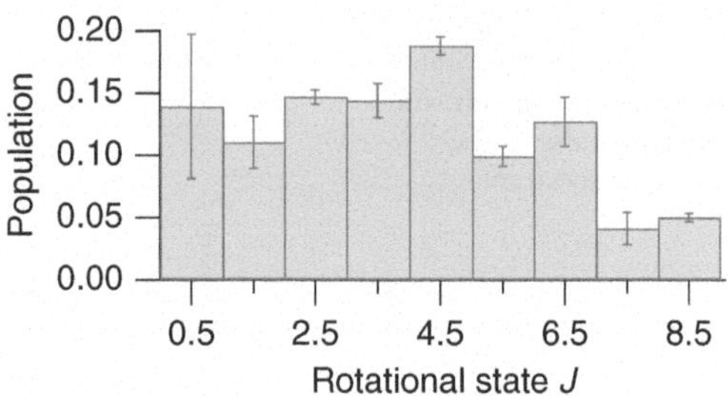

Fig. 3.4 The rotational population of the wave packet extracted from the REMPI spectrum in Fig. 3.3b. The rotational state J has a half-integer due to an unpaired electron of NO

the selection rules of $\Delta J = \pm 1, \pm 2$. However, the interaction term of $\Delta J = \pm 2$ becomes superior to that of $\Delta J = \pm 1$ when J increases. As a result, after the population transfer from the initial $J = 0.5$ to $J = 1.5$ and $J = 2.5$, $J = 1.5$ and $J = 2.5$ states are predominantly coupled with $J = 3.5$ and $J = 4.5$, respectively, due to the main contribution of $\Delta J = \pm 2$. Finally, two pathways of $J = 0.5 \rightarrow 1.5 \rightarrow 3.5 \rightarrow 5.5 \rightarrow \cdots$ and $J = 0.5 \rightarrow 2.5 \rightarrow 4.5 \rightarrow 6.5 \rightarrow \cdots$ contribute the rotational excitation induced by the intense laser fields. The time dependent population calculation supports this explanation [20].

3.3.3 Relative Phase of the Wave Packet

As mentioned in the previous section, a rotational population of a wave packet can be measured with spectroscopic techniques. The information on the rotational population is, however, not enough to reconstruct the wave packet because the wave packet is completely determined by knowledge of both amplitudes A_J and phases δ_J of the expansion coefficients $c_J = A_J e^{i\delta_J}$ in (3.10). Here A_J and δ_J are real numbers. In order to reconstruct the wave packet, a simultaneous measurement of relative phases as well as rotational population, A_J^2, are required.

An experiment to determine the rotational wave packet has been carried out [23]. In the experiment, supersonic molecular beam of a mixture of benzene and helium with a stagnation pressure of 90 bar is used as sample gas. A pump pulse called a 1st pump pulse (70 fs, 820 nm, 1 kHz, 8.4 TW/cm^2) is irradiated to the sample gas and creates the rotational wave packet of benzene. After the delay of τ, a replica of the 1st pump called a 2nd pump pulse is interacted with the wave packet and modulates the rotational population. After 100 ns, a second harmonic of a nanosecond laser pulse called a probe pulse ($\sim$258 nm, an energy resolution of 0.05 cm^{-1}, 10 ns, $\leq$10 μJ/pulse) ionizes benzene molecules by (1 + 1) REMPI via the S_1–S_0 6_0^1 band. By monitoring the ion yield against the probe wavelength, the excitation spectra are measured. In order to obtain phases of the wave packet, the generated $C_6H_6^+$ ion yield at the fixed wavelength of the probe pulse is measured as a function of the delay τ between the 1st pump and the 2nd pump.

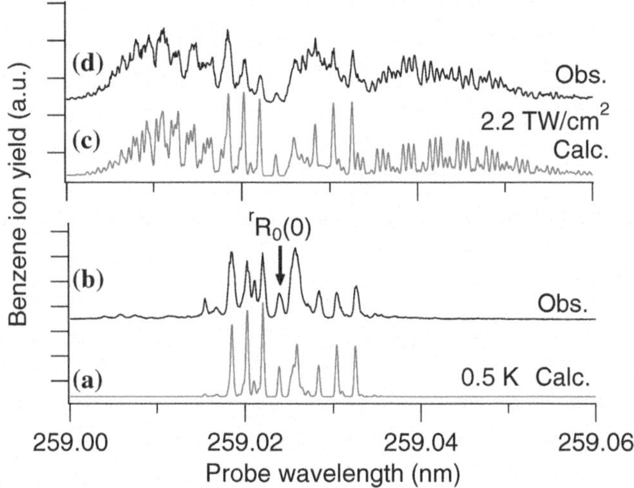

Fig. 3.5 The calculated REMPI spectra of benzene **a** without the pump pulse and **c** with the pump pulse. The observed REMPI spectra of benzene measured **b** without the pump pulse and **d** with the pump pulse

Figure 3.5a, b are the calculated and observed REMPI spectra without the 1st and the 2nd pump. The calculation was performed by assuming the Boltzmann distribution of rotational temperature of 0.5 K and considering transition probabilities of the S_1-S_0 6_0^1 band with the laser energy resolution of 0.05 cm^{-1}. By comparing these spectra, it is found that the rotational temperature of benzene is achieved to be less than 0.5 K with the aid of a higher stagnation pressure of 90 bar. Since benzene is classified as (oblate) symmetric top, an rotational eigen state is described as $|J, K, M\rangle$, where J is the rotational angular momentum, and K and M are its projections onto the molecular symmetry axis and the space fixed axis parallel to the laser polarization direction, respectively. In the case of our experimental condition, the initial rotational states almost restricted to levels with $J_{|K|} = 0_0, 1_0, 1_1, 2_2$ and 3_3, which belong to different irreducible representations of nuclear spin functions.

Figure 3.5c, d are the calculated and observed REMPI spectra with the 1st pump pulse with the light intensity of 2.2 TW/cm^2 and the pulse width stretched to 700 fs so as to avoid ionization and align molecules efficiently. The 2nd pump pulse was not irradiated. The spectral lines emerge over wide wavelength range from 259.00 to 259.06 nm. The rotational excitation up to $J = 10$ are identified by comparing with the calculation.

Figure 3.6 shows the $C_6H_6^+$ ion yield dependence on the delay τ measured with the ${}^rR_0(0)$ transition (from the $J = 0, K = 0$ rotational level in the S_0 state to the $J = 1, K = 1$ one in the S_1 state) as marked by an arrow in Fig. 3.5b. The ion yield is normalized by the ion yield generated by the probe pulse without the 1st and 2nd pump.

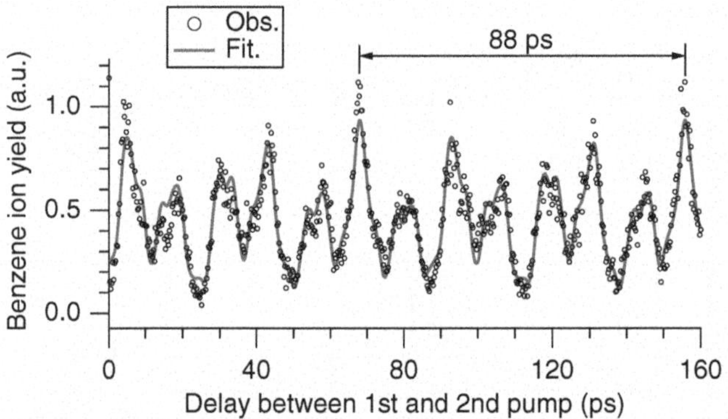

Fig. 3.6 The dependence of the benzene ion yield measured with the $r_0^R(0)$ transition, which corresponds to population of the initial state $J = 0, M = 0$, on the delay between the 1st and the 2nd pump

The observed signal includes the information on the amplitudes A_J and phases δ_J as explained below. The initial rotational state, $|J_i, K_i, M_i\rangle = |0, 0, 0\rangle$, irradiated by the 1st pump pulse at the time $t = 0$, changes into the wave packet $|\Psi(t)\rangle$,

$$|\Psi(t)\rangle = \hat{U}(t, 0)|0, 0, 0\rangle = \sum_J A_J e^{i\delta_J} e^{-i\omega_{J,0}t}|J, 0, 0\rangle, \tag{3.16}$$

where $\hat{U}(t_2, t_1)$ is the time evolution operator from time t_1 to t_2, $\omega_{J,K} = E_{J,K}/\hbar$, $E_{J,K} = hcBJ(J+1) + hc(C-B)K^2$ represents an eigen energy of an eigen state $|J, K, M\rangle$, and $B = 0.1897717\,\text{cm}^{-1}$ and $C = B/2 = 0.0948859\,\text{cm}^{-1}$ are rotational constants [94]. In (3.16), the summation isn't taken over K and M because the initial K and M values are conserved in the case of the linear polarization. If the 2nd pump pulse is irradiated at $t = \tau$, the wave packet expressed by (3.16) is changed into

$$|\Psi(t)\rangle = \hat{U}(t, \tau)|\Psi(\tau)\rangle = \sum_J B_J(\tau) e^{-i\omega_{J,0}t}|J, 0, 0\rangle, \tag{3.17}$$

where $B_J(\tau)$ is the transition amplitude from the initial $|0, 0, 0\rangle$ to $|J, 0, 0\rangle$. Since the observed signal is proportional to the population of the $|0, 0, 0\rangle$ state, which is not only the initial state but also the final probed state, the observed signal are represented by $|B_0(\tau)|^2$. The population can be analytically given by [23]

$$|B_0(\tau)|^2 = \sum_{J'} A_{J'}^4 + 2\sum_{J'>J''} A_{J'}^2 A_{J''}^2 \cos\left\{(\omega_{J'0} - \omega_{J''0})\,\tau + 2\,(\delta_{J'} - \delta_{J''})\right\}. \tag{3.18}$$

Consequently, the ion yield generated by the double pump and population probe method includes the information on both amplitudes and phases. The observed ion

Table 3.1 The population and phase of the rotational wave packet of benzene irradiated by a intense laser fields of 8.4 TW/cm^2 and 70 fs

J	0	2	4	6	8
Population	0.1	0.65	0.19	0.052	0.0065
Phase	0	−2.2	−4.0	−5.6	−7.3

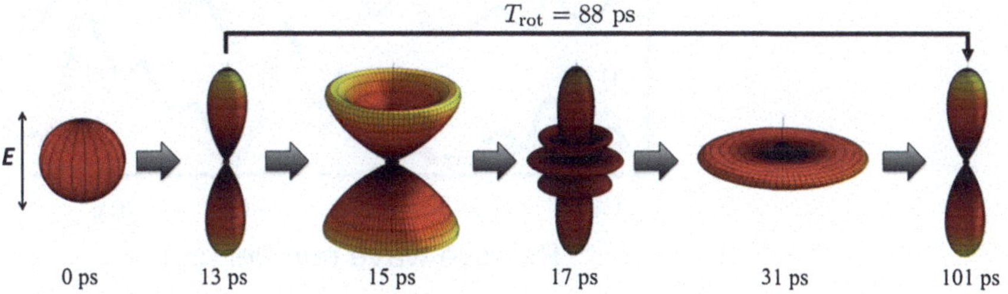

Fig. 3.7 The time evolution of the reconstracted 3D rotational wave packet of benzene

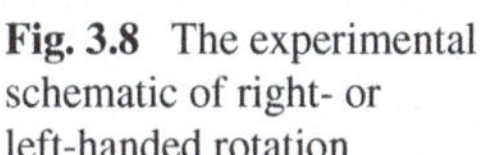

Fig. 3.8 The experimental schematic of right- or left-handed rotation

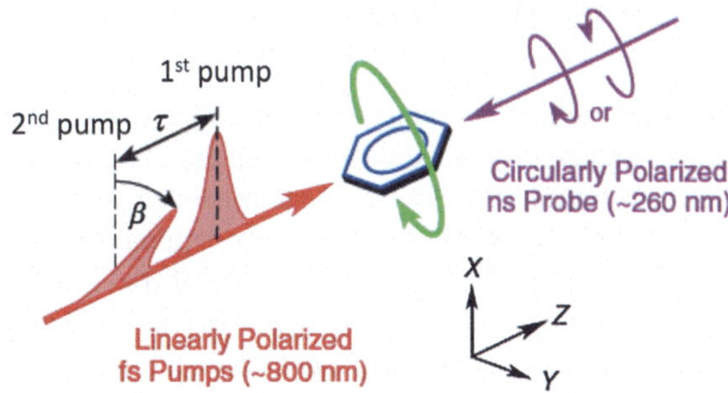

yield is fitted well by (3.18) as shown in Fig. 3.6. The determined fitting parameters are listed in Table 3.1.

The dynamics of the wave packet, that is, the molecular axis distribution, is traced by using the obtained parameters. Figure 3.7 shows the time evolution of the reconstructed wave packet. The axis distribution of the initial $|0, 0, 0\rangle$ state is isotropic before the laser pulse is irradiated at $t = 0$. After the interaction, the sharp distribution parallel to the polarization direction, i.e., nonadiabatic molecular alignment, is achieved at $t = 13$ ps. Then, at $t = 31$ ps, a doughnut-shape distribution perpendicular to the polarization direction, i.e., anti-alignment, is achieved. We found that the localized distribution along the polarization direction at $t = 101$ ps is the same as that at $t = 13$ ps. The time difference of 88 ps between 13 and 101 ps equals to the rotational period of benzene, $T_{rot} = \frac{1}{2Bc} = 88$ ps.

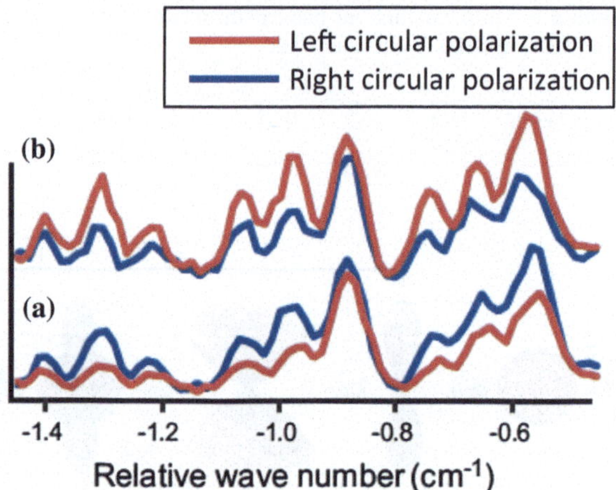

Fig. 3.9 The REMPI spectra of **a** right- or **b** left-handed rotating molecules measured with the probe pulses with right- or left-circular polarization

3.3.4 The Left-Handed and Right-Handed Rotational Motion of the Wave Packet

The wave packet motion as mentioned above doesn't correspond to a classical left-handed and right-handed rotation. Kitano et al. has demonstrated that a molecular axis distribution rotated like a right-handed pinwheel or a left-handed pinwheel [24, 95, 96]. Since the direction of rotational motion is expressed by a sign of a quantum number M, right- or left-handed rotational motion can be induced if the population between positive M states and negative M states doesn't balance.

In order to make the wave packet with a unbalanced M distribution, two delayed intense laser pulses with skewed polarization directions are used. The first pulse with a linear polarization along X-axis propagates along Z-axis. The second pulse with a linear polarization tilted by the angle, β, from X-axis also propagates along Z-axis as shown in Fig. 3.8. The 1st pulse creates a rotational wave packet. After a certain time, the molecular axis distribution concentrates along the polarization direction of the 1st pulse, i.e., NAMA takes place. The 2nd pulse irradiated at this instant gives a torque to align along the polarization direction of the 2nd pulse. Consequently, molecules rotate toward the 1st- to the 2nd-polarization direction. By changing a sign of β, the left- and right-handed rotational motion is controlled.

The direction of rotational motion is probed by a nanosecond laser pulse with a circular polarization. The absorption probability of right(left)-handed rotating molecules differs between a right- and a left-circular polarized radiation, that is, these molecules exhibit circular dichroism.

Figure 3.9a, b show the REMPI spectra for right- and left-handed rotating molecules created by using $\beta = 45°$ and $-45°$, respectively. The spectra show the circular dichroism. It is clear evidence for the generation of right- and left-handed rotating wave packet.

3.4 Summary

In summary, the nonadiabatic molecular alignment (NAMA) and the nonadiabatic molecular orientation (NAMO) induced by intense nonresonant short laser pulses are introduced. The reconstruction of the rotational wave packet as well as the population measurement has been demonstrated by a state-selective probe using a spectroscopic technique. We also demonstrate the creation of the peculiar rotational wave packet corresponding to right- or left-handed rotation.

In future, a creation and control of a vibrational wave packet generated by the same method as the NAMA is expected because of the dependence of the anisotropic polarizability on vibrational motion. Recently, the vibrational wave packet in the electronic ground state generated by a intense nonresonant femtosecond laser pulse has been reported [97–100]. One of the best target for the creation of the vibrational wave packet by using an intense laser field is weakly-bounded molecular cluster due to having lower vibrational frequencies and large amplitude motion. We have already started to investigate these cluster target such as NO-Ar cluster and benzene clusters and succeeded in the observation of the vibrational wave packet.

References

1. H. Stapelfeldt, T. Seideman, Aligning molecules with strong laser pulses. Rev. Mod. Phys. **75**(2), 543–557 (2003)
2. T. Seideman, E. Hamilton, Nonadiabatic alignment by intense pulses. concepts, theory, and directions. Advances in Atomic, Molecular and Optical. Physics **52**, 289–329 (2005)
3. Y. Ohshima, H. Hasegawa, Coherent rotational excitation by intense nonresonant laser fields. Int. Rev. Phys. Chem. **29**(4), 619–663 (Oct–Dec 2010)
4. F. Rosca-Pruna, M.J.J. Vrakking, Experimental observation of revival structures in picosecond laser-induced alignment of I_2. Phys. Rev. Lett. **87**, 153902 (4 pages) (2001)
5. F. Rosca-Pruna, M.J.J. Vrakking, Revival structures in picosecond laser-induced alignment of I_2 molecules. i. experimental resutls. J. Chem. Phys. **116**, 6567–6578 (2002)
6. P.W. Dooley, I.V. Litvinyuk, Kevin F. Lee, D.M. Rayner, M. Spanner, D.M. Villeneuve, P.B. Corkum, Direct imaging of rotational wave-packet dynamics of diatomic molecules. Phys. Rev. A **68**, 023406 (12 pages) (2003)
7. C.H. Lin, J.P. Heritage, T.K. Gustafson, R.Y. Chiao, J.P. McTague, Birefringence arising from the reorientation of the polarizability anisotropy of molecules in collisionless gases. Phys. Rev. A **13**, 813–829 (1976)
8. V. Renard, M. Renard, S. Guérin, Y.T. Pashayan, B. Lavorel, O. Faucher, H.R. Jauslin, Post-pulse molecular alignment measured by a weak field polarization technique. Phys. Rev. Lett. **90**, 153601 (4 pages) (2003)
9. O. Faucher, B. Lavorel, E. Hertz, F. Chaussard, Progress in ultrafast intense laser science vii. pp. 79–108 (2011)
10. E.T. McCole, J.H. Odhner, D.A. Romanov, R.J. Levis, Spectral-to-temporal amplitude mapping polarization spectroscopy of rotational transients. J. Phys. Chem. A **117**, 6354–6361 (2013)
11. J. Itatani, D. Zeidler, J. Levesque, M. Spanner, D.M. Villeneuve, P.B. Corkum, Controlling high harmonic generation with molecular wave packets. Phys. Rev. Lett **94**, 123902 (4 pages) (2005)

12. K. Miyazaki, M. Kaku, G. Miyaji, A. Abdurrouf, F.H.M. Faisal, Field-free alignment of molecules observed with high-order harmonic generation. Phys. Rev. Lett. **95**, 243903 (4 pages) (2005)
13. T. Kanai, S. Minemoto, H. Sakai, Quantum interference during high-order harmonic generation from aligned molecules. Nature **435**, 470–474 (2005)
14. T. Kanai, S. Minemoto, H. Sakai, Ellipticity dependence of high-order harmonic generation from aligned molecules. Phys. Rev. Lett. **98**, 053002 (4 pages) (2007)
15. X. Guo, P. Liu, Z. Zeng, Y. Pengfei, R. Li, X. Zhizhan, Ionization effects on field-free molecular alignment observed with high-order harmonic generation. Optics Commun. **282**, 2539–2542 (2009)
16. V.G. Stavros, E. Harel, S.R. Leone, The influence of intense control laser pulses on homodyne-detected rotational wave packet dynamics in O_2 by degenerate four-wave mixing. J. Chem. Phys. **122**, 064301 (9 pages) (2005)
17. S. Fleischer, I.Sh. Averbukh, Y. Prior, Isotope-selective laser molecular alignment. Phys. Rev. A **74**, 041403(R) (4 pages) (2006)
18. S. Fleischer, I.Sh. Averbukh, Y. Prior, Selective alignment of molecular spin isomers. Phys. Rev. Lett. **99**, 093002 (4 pages) (2007)
19. X. Ren, V. Makhija, V. Kumarappan, Measurement of field-free alignment of jet-cooled molecules by nonresonant femtosecond degenerate four-wave mixing. Phys. Rev. A **85** 033405 (6 pages) (2012)
20. H. Hasegawa, Y. Ohshima, Decoding the state distribution in a nonadiabatic rotational excitation by a nonresonant intense laser field. Phys. Rev. A **74**, 061401(R) (4 pages) (2006)
21. A.S. Meijer, Y. Zhang, D.H. Parker, W.J. van der Zande, A. Gijsbertsen, M.J.J. Vrakking, Controlling rotational state distributions using two-pulse stimulated raman excitation. Phys. Rev. A **76**, 023411 (9 pages) (2007)
22. H. Hasegawa, Y. Ohshima, Nonadiabatic rotational excitation of benzene by nonresonant intense femtosecond laser fields. Chem. Phys. Lett. **454**, 148–152 (2008)
23. H. Hasegawa, Y. Ohshima, Quantum state reconstruction of a rotational wave packet created by a nonresonant intense femtosecond laser field. Phys. Rev. Lett. **101** 053002 (4 pages) (2008)
24. K. Kitano, H. Hasegawa, Y. Ohshima, Ultrafast angular momentum orientation by linearly polarized laser fields. Phys. Rev. Lett. **103**, 223002 (4 pages) (2009)
25. R.O. Ghafur, K. Vidma, A. Gijsbertsen, O.M. Shir, T. Bäck, A. Meijer, W.J. van der Zande, D. Parker, M.J.J. Vrakking, Evolutionary optimization of rotational population transfer. Phys. Rev. A **84**, 033415 (5 pages) (2011)
26. D. Baek, H. Hasegawa, Y. Ohshima, Unveiling the nonadiabatic rotational excitation process in a symmetric-top molecule induced by two intense laser pulses. J. Chem. Phys. **134**, 224302 (10 pages) (2011)
27. T. Seideman, Rotational excitation and molecular alignment in intense laser fields. J. Chem. Phys. **103**, 7887–7896 (1995)
28. T. Seideman, New means of spatially manipulating molecules with light. J. Chem. Phys. **111**, 4397–4405 (1999)
29. T. Seideman, Revival structure of aligned rotational wave packets. Phys. Rev. Lett. **83**, 4971–4974 (1999)
30. J. Ortigoso, M. Rodríguez, M. Gupta, B. Friedrich, Time evolution of pendular states created by the interaction of molecular polarizability with a pulsed nonresonant laser field. J. Chem. Phys. **110**, 3870–3875 (1999)
31. T. Seideman, On the dynamics of rotationally broad, spatially aligned wave packets. J. Chem. Phys. **115**, 5965–5973 (2001)
32. I.Sh. Averbukh, R. Arvieu, Angular focusing, squeezing, and rainbow formation in a strongly driven quantum rotor. Phys. Rev. Lett. **87**, 163601 (4 pages) (2001)
33. M. Leibscher, I.Sh. Averbuck, H. Rabitz, Molecular alignment by trains of short laser pulses. Phys. Rev. Lett. **90**, 213001 (4 pages) (2003)

34. M. Spanner, E.A. Shapiro, M. Ivanov, Coherent control of rotational wave-packet dynamics via fractional revivals. Phys. Rev. Lett. **92**, 093001 (4 pages) (2004)
35. O.M. Shir, V. Beltrani, Th. Bäck, H. Rabitz, M.J.J. Vrakking, On the diversity of multiple optimal controls for quantum systems. J. Phys. B **41**, 074021 (11 pages) (2008)
36. K. Nakagami, Y. Mizumoto, Y. Ohtsuki, Optimal alignment control of a nonpolar molecule through nonresonant multiphoton transitions. J. Chem. Phys. **129**, 194103 (9 pages) (2008)
37. E. Hamilton, T. Seideman, T. Ejdrup, M.D. Poulsen, C.Z .Bisgaard, S.S. Viftrup, H. Stapelfeldt, Alignment of symmetric top molecules by short laser pulses. Phys. Rev. A **72**, 043402 (12 pages) (2005)
38. C.Z. Bisgaard, S.S. Viftrup, Alignment enhancement of a symmetric top molecule by two short laser pulses. Phys. Rev. A **73**, 053410 (9 pages) (2006)
39. E. Péronne, M.D. Poulsen, C.Z. Bisgaard, H. Stapelfeldt, T. Seideman, Nonadiabatic alignment of asymmetric top molecules: field-free alignment of iodobenzene. Phys. Rev. Lett. **91**, 043003 (4 pages) (2003)
40. E. Péronne, M.D. Poulsen, H. Stapelfeldt, C.Z. Bisgaard, E. Hamilton, T. Seideman, Nonadiabatic laser-induced alignment of iodobenzene molecules. Phys. Rev. A **70**, 063410 (9 pages) (2004)
41. J.G. Underwood, B.J. Sussman, A. Stolow, Field-free three dimensional molecular axis alignment. Phys. Rev. Lett. **94**, 143002 (4 pages) (2005)
42. K.F. Lee, D.M. Villeneuve, P.B. Corkum, A. Stolow, J.G. Underwood, Field-free three-dimensional alignment of polyatomic molecules. Phys. Rev. Lett. **97**, 173001 (4 pages) (2006)
43. A. Rouzée, S. Guérin, V. Boudon, B. Lavorel, O. Faucher, Field-free one-dimensional alignment of ethylene molecule. Phys. Rev. A **73**, 033418 (9 pages) (2006)
44. L. Holmegaard, S.S. Viftrup, V. Kumarappan, C.Z. Bisgaard, H. Stapelfeldt, Control of rotational wave-packet dynamics in asymmetric top molecules. Phys. Rev. A **75** 051403(R) (4 pages) (2007)
45. S.S. Viftrup, V. Kumarappan, L. Holmegaard, C.Z. Bisgaard, H. Stapelfeldt, M. Artamonov, E. Hamilton, T. Seideman, Controlling the rotation of asymmetric top molecules by the combination of a long and a short laser pulse. Phys. Rev. A **79**, 023404 (12 pages) (2009)
46. V. Makhija, X. Ren, V. Kumarappan, Metric for three-dimensional alignment of molecules. Phys. Rev. A **85**, 033425 (6 pages) (2012)
47. L.S. Spector, M. Artamonov, S. Miyabe, T. Martinez, T. Seideman, M. Guehr, P.H. Bucksbaum, Axis-dependence of molecular high harmonic emission in three dimensions. Nat. Commun. **5**, 3190 (2014) doi:10.1038/ncomms4190
48. J. Wu, A. Vredenborg, B. Ulrich, L.Ph.H. Schmidt, M. Meckel, S. Voss, H. Sann, H. Kim, T. Jahnke, R. Dörner, Nonadiabatic alignment of van der waals-force-bound argon dimers by femtosecond laser pulses. Phys. Rev. A **83**, 061403(R) (4 pages) (2011)
49. A. von Veltheim, B. Borchers, G. Steinmeyer, H. Rottke, Imaging the impulsive alignment of noble-gas dimers via Coulomb explosion. Phys. Rev. A **89**, 023432 (12 pages) (2014)
50. G. Galinis, L.G.M. Luna, M.J. Watkins, A.M. Ellis, R.S. Minns, M. Mladenović, M. Lewerenz, R.T. Chapman, I.C.E. Turcu, C. Cacho, E. Springate, L. Kazak, S. Göde, R. Irsig, S. Skruszewicz, J. Tiggesbäumker, K.-H. Meiwes-Broer, A. Rouzée, J.G. Underwood, M. Siano, K. von Haeften, Formation of coherent rotational wavepackets in small molecule-helium clusters using impulsive alignment. Faraday Discuss. **171**, 195–218 (2014)
51. K.F. Lee, I.V. Litvinyuk, P.W. Dooley, M.Spanner, D.M. Villeneuve, P.B. Corkum, Two-pulse alignment of molecules. J. Phys. B **37**, L43–L48 (2004)
52. M. Leibscher, I.Sh. Averbukh, H. Rabitz, Enhanced molecular alignment by short laser pulses. Phys. Rev. A **69**, 013402 (10 pages) (2004)
53. C.Z. Bisgaard, M.D. Poulsen, E. Péronne, S.S. Viftrup, H. Stapelfeldt, Observation of enhanced field-free molecular alignment by two laser pulses. Phys. Rev. Lett. **92** 173004 (4 pages) (2004)
54. J.P. Cryan, P.H. Bucksbaum, R.N. Coffee, Field-free alignment in repetitively kicked nitrogen gas. Phys. Rev. A **80**, 063412 (5 pages) (2009)

55. S. Zhdanovich, A.A. Milner, C. Bloomquist, J. Fuoß, I.Sh. Averbuck, J.W. Hepburn, V. Milner, Control of molecular rotation with a chiral train of ultrashort pulses. Phys. Rev. Lett. **107**, 243004 (5 pages) (2011)
56. M. Renard, E. Hertz, B. Lavorel, O. Faucher, Controlling ground-state rotational dynamics of molecules by shaped femtosecond laser pulses. Phys. Rev. A **69**, 043401 (6 pages) (2004)
57. M. Renard, E. Hertz, S. Guérin, H. R.Jauslin, B. Lavorel, O. Faucher, Control of field-free molecular alignment by phase-shaped laser pulses. Phys. Rev. A **72**, 025401 (4 pages) (2005)
58. C. Horn, M. Wollenhaupt, M. Krug, T. Baumert, R. de Nalda, L. Bañares, Adaptive control of molecular alignment. Phys. Rev. A **73**, 031401(R) (4 pages) (2006)
59. E. Hertz, A. Rouzée, S. Guérin, B. Lavorel, O. Faucher, Optimization of field-free molecular alignment by phase-shaped laser pulses. Phys. Rev. A **75**, 031403(R) (4 pages) (2007)
60. T. Suzuki, Y. Sugawara, S. Minemoto, H. Sakai, Optimal control of nonadiabatic alignment of rotationally cold N_2 molecules with the feedback of degree of alignment. Phys. Rev. Lett. **100**, 033603 (6 pages) (2008)
61. H. Akagi, H. Ohba, K. Yokoyama, A. Yokoyama, K. Egashira, Y. Fujimura, Rotational-coherence molecular laser isotope separation. Appl. Phys. B **95**, 17–21 (2009)
62. H. Akagi, T. Kasajima, T. Kumada, R. Itakura, A. Yokoyama, H. Hasegawa, Y. Ohshima, Isotope-selective ionization utilizing molecular alignment and non-resonant multiphoton ionization. Appl. Phys. B **109**, 75–80 (2012)
63. I.V. Litvinyuk, K.F. Lee, P.W. Dooley, D.M. Rayner, D.M. Villeneuve, P.B. Corkum, Alignment-dependent strong field ionization of molecules. Phys. Rev. Lett. **90** 233003 (4 pages) (2003)
64. D. Pinkham, R.R. Jones, Intense laser ionization of transiently aligned CO. Phys. Rev. A **72**, 023418 (6 pages) (2005)
65. D. Pavičić, K.F. Lee, D.M. Rayner, P.B. Corkum, D.M. Villeneuve, Direct measurement of the angular dependence of ionization for N_2 in intense laser fields. Phys. Rev. Lett. **98**, 243001 (4 pages) (2007)
66. V. Kumarappan, L. Holmegaard, C. Martiny, C.B. Madsen, T.K. Kjeldsen, S.S. Viftrup, L.B. Madsen, H. Stapelfeldt, Multiphoton electron angular distributions from laser-aligned CS_2 molecules. Phys. Rev. Lett. **100**, 093006 (4 pages) (2008)
67. A.-T. Le, R.R. Lucchese, M.T. Lee, C.D. Lin, Probing molecular frame photoionizaiton via laser generated high-order harmonics from aligned molecules. Phys. Rev. Lett. **102**, 203001 (4 pages) (2009)
68. R. Itakura, H. Hasegawa, Y. Kurosaki, A. Yokoyama, Y. Ohshima, Coherent correlation between nonadiabatic rotational excitation and angle-dependent ionization of NO in intense laser fields. J. Phys. Chem. A **114**, 11202–11209 (2010)
69. F. Kelkensberg, A. Rouzée, W. Siu, G. Gademann, P. Johnsson, M. Lucchini, R.R. Lucchese, M.J.J. Vrakking, XUV ionization of aligned molecules. Phys. Rev. A **83**, 023406 (11 pages) (2011)
70. J. Itatani, J. Levesque, D. Zeidler, H. Niikura, H. Pépin, J.C. Kieffer, P.B. Corkum, D.M. Villeneuve, Tomographic imaging of molecular orbitals. Nature **432**, 867–871 (2004)
71. S. Minemoto, T. Kanai, H. Sakai, Alignment dependence of the structural deformation of CO_2 molecules in an intense femtosecond laser field. Phys. Rev. A **77**, 041401(R) (4 pages) (2008)
72. X. Xie, K. Doblhoff-Dier, H. Xu, S. Roither, M.S. Schöffler, D. Kartashov, S. Erattupuzha, T. Rathje, G.G. Paulus, K. Yamanouchi, A. Baltuška, S. Gräfe, M. Kitzler, Selective control over fragmentation reactions in polyatomic molecules using impulsive laser alignment. Phys. Rev. Lett. **112**, 163003 (5 pages) (2014)
73. P. Reckenthaeler, M. Centurion, W. Fuß, S.A. Trushin, F. Krausz, E.E. Fill, Time-resolved electron diffraction from selectively aligned molecules. Phys. Rev. Lett. **102**, 213001 (4 pages) (2009)
74. S. Fleischer, Y. Zhou, R.W. Field, K.A. Nelson, Molecular orientation and alignment by intense single-cycle THz pulses. Phys. Rev. Lett. **107**, 163603 (5 pages) (2011)

75. K. Kitano, N. Ishii, J. Itatani, High degree of molecular orientation by a combination of THz and femtosecond laser pulses. Phys. Rev. A **84**, 053408 (7 pages) (2011)
76. K. Kitano, N. Ishii, N. Kanda, Y. Matsumoto, T. Kanai, M. Kuwata-Gonokami, J. Itatani, Orientation of jet-cooled polar molecules with an intense single-cycle THz pulse. Phys. Rev. A **88**, 061405(R) (5 pages) (2013)
77. K.N. Egodapitiya, S. Li, R.R. Jones, Terahertz-induced field-free orientation of rotationally excited molecules. Phys. Rev. Lett. **112**, 103002 (5 pages) (2014)
78. A. Goban, S. Minemoto, H. Sakai, Laser-field-free molecular orientation. Phys. Rev. Lett. **101**, 013001 (4 pages) (2008)
79. S. De, I. Znakovskaya, D. Ray, F. Anis, N.G. Johnson, I.A. Bocharova, M. Magrakvelidze, B.D. Esry, C.L. Cocke, I.V. Litvinyuk, M.F. Kling, Field-free orientation of CO molecules by femtosecond two-color laser fields. Phys. Rev. Lett. **103**, 153002 (4 pages) (2009)
80. P.M. Kraus, D. Baykusheva, H.J. Wörner, Two-pulse field-free orientation reveals anisotropy of molecular shape resonance. Phys. Rev. Lett. **113**, 023001 (5 pages) (2014)
81. *Special issue of Stereodynamics of Chemical Reaction*, vol. 101 (1997)
82. C.Z. Bisgaard, O.J. Clarkin, W. Guorong, A.M.D. Lee, O. Geßner, C.C. Hayden, A. Stolow, Time-resolved molecular frame dynamics of fixed-in-space CS_2 molecules. Science **323**, 1464–1468 (2009)
83. Tsuneto Kanai, Hirofumi Sakai, Numerical simulations of molecular orientation using strong, nonresonant, two-color laser fields. J. Chem. Phys. **115**(12), 5492–5497 (2001)
84. E. Hertz, D. Daems, S. Guérin, H.R. Jauslin, B. Lavorel, O. Faucher, Field-free molecular alignment induced by elliptically polarized laser pulses: Noninvasive three-dimensional characterization. Phys. Rev. A **76**, 043423 (5 pages) (2007)
85. C.T.L. Smeenk, P.B. Corkum, Molecular alignment using circularly polarized laser pulses. J. Phys. B **46**, 201001 (4 pages) (2013)
86. K. Oda, M. Hita, S. Minemoto, H. Sakai, All-optical molecular orientation. Phys. Rev. Lett. **104**, 213901 (4 pages) (2010)
87. R.N. Zare, *Angular Momentum*. Wiley, New York (1988)
88. K.P. Huber, G. Herzberg, *Constants of Diatomic Molecules* (data prepared by J.W. Gallagher, R.D. Johnson, III) in NIST Chemistry WebBook, NIST Standard Reference Database Number 69, eds. by P.J. Linstrom, W.G. Mallard, National Institute of Standards and Technology, Gaithersburg MD, 20899, http://webbook.nist.gov
89. F.J. Lovas, Microwave spectral tables ii. triatomic molecules. J. Phys. Chem. Ref. Data **7**, 1445–1750 (1978)
90. S. Zhao, P. Liu, R. Li, X. Zhizhan, Controlling coherent population transfer in molecular alignment using two laser pulses. Chem. Phys. Lett. **480**, 67–70 (2009)
91. C.H. Lin, J.P. Heritage, T.K. Gustafsonn, Susceptibility echos in linear molecular gases. Appl. Phys. Lett. **19**, 397–1302 (1971)
92. J.P. Heritage, T.K. Gustafson, C.H. Lin, Observation of coherent transient birefringence in CS_2 vapor. Phys. Rev. Lett. **34**, 1299–1302 (1975)
93. F. Rosca-Pruna, M.J.J. Vrakking, Revival structures in picosecond laser-induced alignment of I_2 molecules. ii. numerical modeling. J. Chem. Phys. **116**, 6579–6588 (2002)
94. A. Doi, M. Baba, S. Kasahara, H. Kato, Sub-doppler rotationally resolved spectroscopy of the $S_1^1 b_{2u}(v_6 = 1) \leftarrow S_0^1 a_{1g}(v = 0)$. J. Mol. Spectrosc. **227**, 180–186 (2004)
95. S. Fleischer, Y. Khodorkovsky, Y. Prior, I.Sh. Averbukh, Controlling the sense of molecular rotation. New J. Phys. **11**, 105039 (15 pages) (2009)
96. Y. Khodorkovsky, K. Kitano, H. Hasegawa, Y. Ohshima, I.Sh. Averbukh, Controlling the sense of molecular rotation: classical versus quantum analysis. Phys. Rev. A **83**, 023423 (10 pages) (2011)
97. N.L. Wagner, A. Wüest, I.P. Christov, T. Popmintchev, X. Zhou, M.M. Murnane, H.C. Kapteyn, Monitoring molecular dynamics using coherent electrons from high harmonic generation. Proc. Natl. Acad. Sci. **103**, 13279–13285 (2006)
98. W. Li, X. Zhou, R. Lock, S. Patchkovskii, A. Stolow, H.C. Kapteyn, M.M. Murnane, Time-resolved dynamics in N_2 probed using high harmonic generation. Science **322**, 1207–1211 (2008)

99. C.B. Madsen, L.B. Madsen, S.S. Viftrup, M.P. Johansson, T.B. Poulsen, L. Holmegaard, V. Kumarappan, K.A. Jørgensen, H. Stapelfeldt, A combined experimental and theoretical study on realizing and using laser controlled torsion of molecules. J. Chem. Phys. **130**, 234310 (9 pages) (2009)
100. J.L. Hansen, J.H. Nielsen, C.B. Madsen, A.T. Lindhardt, M.P. Johansson, T. Skrydstrup, L.B. Madsen, H. Stapelfeldt, Control and femtosecond time-resolved imaging of torsion in a chiral molecule. J. Chem. Phys. **136**, 204310 (10 pages) (2012)

Chapter 4
Dynamics of Atomic Clusters Under Intense Femtosecond Laser Pulses

Gaurav Mishra and N.K. Gupta

Abstract This chapter presents a review of theoretical modelling of intense ultra-short laser interaction with atomic clusters. We start with a brief description of interaction of laser with gas atoms and solid targets. This is followed by a general introduction of the non-linear, non-perturbative interaction of strong laser fields with atomic clusters. Various theoretical models are presented to explain the rich features observed in laser irradiated clusters. A molecular dynamic model developed by us to study laser-cluster interaction is briefly described. Various problems of contemporary interest like anisotropic emission of ions from argon and xenon clusters, effect of carrier-envelope phase on ionization dynamic of xenon clusters, neutron emission from deuterium clusters etc. are investigated and main results of these studies are presented.

4.1 Introduction

The development of chirped pulse amplification (CPA) [1, 2] and Kerr-lens mode-locking [3] has enabled researchers to generate routinely strong ($>10^{14}$ W/cm^2) and short (<1 ps) laser pulses. This invention has revived the field of laser matter interaction. At this intensity, the electric field of the laser pulse becomes comparable to the atomic unit of electric field strength ($\mathcal{E}_a$)[1] that leads to many interesting features. One should note that the way light couples to matter is a strong function of incident laser intensity. At low light intensities, the ionization of matter can be described adequately by Einstein's law of photoelectric effect where a single photon absorption leads to emission of an electron. When intensity of light is increased up to

[1] $\mathcal{E}_a$ is the strength of the Coulomb field experienced by an electron in the first Bohr orbit of the hydrogen atom which comes out to be equal to 5.1×10^9 V/cm^{-1}. When expressed in terms of intensity, it is equal to 3.5×10^{16} Wcm^{-2}.

G. Mishra (✉) · N.K. Gupta
Theoretical Physics Division, Bhabha Atomic Research Centre,
Mumbai 400094, India
e-mail: gauravm@barc.gov.in

N.K. Gupta
e-mail: nkgupta@barc.gov.in

K. Yamanouchi et al. (eds.), *Progress in Ultrafast Intense Laser Science XII*,
Springer Series in Chemical Physics 112, DOI 10.1007/978-3-319-23657-5_4

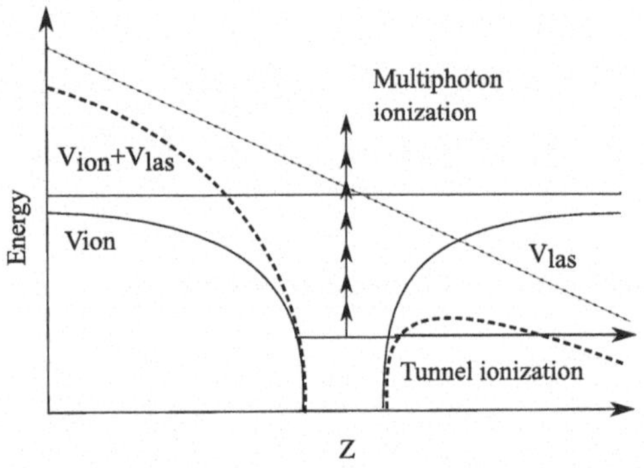

Fig. 4.1 Schematic view of MPI and TI in atoms. $V_{ion} \propto -q/|z|$, $V_{las} \propto -z$ and $V_{ion} + V_{las} \propto -q/|z| - z$ represent the potentials of unperturbed q-charged ion, laser field and their effective sum, respectively

10^{10} W/cm^2, absorption of more than one photon leads to the ejection of a bound electron (Fig. 4.1), commonly known as multi-photon ionization (MPI). The light intensity is still far below the atomic unit of intensity and the processes can still be described perturbatively. At intensities of 10^{13}–10^{14} W/cm^2, the effects of external field can no longer be treated as a perturbation to the coulomb field experienced by an electron in the atom. This leads to the emergence of higher order MPI, tunnell ionization (TI) and over the barrier ionization (OTBI) [4]. In the case of TI, the external field distorts the coulomb barrier experienced by electron in the atom to such a extent that the electron tunnels through the barrier to continuum. If the suppression of barrier is so strong that it goes below the bound state of electron, the electron does not see any barrier. The probability of ionization is unity in this case and the process is known as OTBI. When isolated atoms e.g. gaseous targets are exposed to intense laser light pulses, rapid ionization by the above mentioned mechanisms leads to the generation of above threshold ionization (ATI) [5], high harmonic generation (HHG) [6, 7], molecular alignment and enhanced ionization (of molecules) [8] etc. The atom/molecule density is low enough so that each atoms responds independently to the electric field of incident light.

On the other hand, exposure of high-density targets to intense laser field leads to generation of electrons via rapid ionization which further enhances the absorption of laser light by various mechanisms such as collisional absorption (inverse bremsstrahlung), resonance absorption and parametric instabilities [9]. All these light absorption processes lead to the generation of high temperature plasma. The behaviour of this plasma in the presence of electromagnetic field is a subject of laser matter interaction. Although inertial confinement fusion (ICF) as a source of energy production has remained an active area of research in this field, other applications like production of ultrafast X-ray pulses [10], fast particle source [11, 12] and production of extreme astrophysical plasma state in laboratory [13] have also motivated the research in this area.

In the last decade, atomic/molecular clusters have drawn a worldwide attention to be used as targets (Fig. 4.2) for laser irradiation [14, 15]. These clusters are aggregates of a countable number (10^2–10^7) of particles (atoms or molecules). Depending on

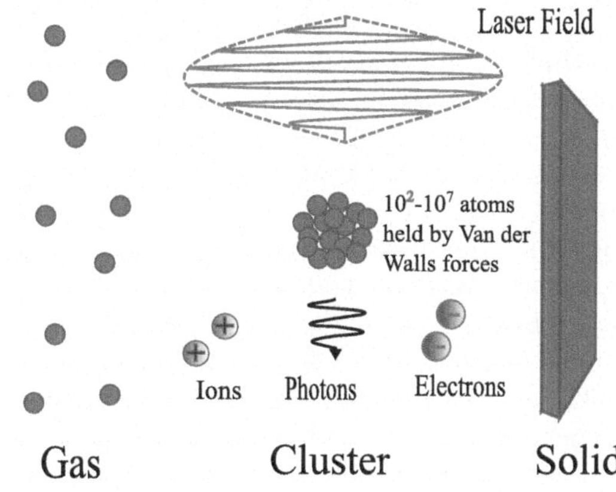

Fig. 4.2 Clusters intermediate to gas and solid target gives rise to the emission of energetic particles after the irradiation by short and intense laser fields

the nature of bonding among the constituent particles, various kinds of cluster (ionic, covalent, metallic or noble gas clusters) can be produced [16]. The study of clusters is mainly motivated by the fact that they occupy a central position between gas and solid. The fundamental question to be answered by these studies is that do the properties of clusters evolve continuously till they form bulk material starting from gaseous phase or do exhibit completely different properties compared to their solid or gas counterpart? The availability of short pulse and intense lasers adds extra feature to the laser excited clusters i.e. one can drive and resolve the ion and electron dynamics on their natural time scales under strong field conditions.

It is important to note that excitation of cluster in the presence of laser field strongly depends upon the nature of the cluster itself. For example, metal clusters already possess the de-localised valence electron cloud [17, 18] whereas rare gas clusters first needs to be metallized under the presence of strong laser fields i.e. the creation of quasi free electrons which can freely roam inside the cluster. The metal clusters undergo collective resonance at much lower laser intensity as compared to their rare gas counterparts [19, 20]. Here we restrict ourselves to noble gas clusters driven by intense femtosecond laser pulses. These targets, intermediate to gases and solids, contain advantages of both the phases. For instance, energy absorption efficiency of laser light by clusters is comparable to solid target plasmas [21] yet the particle emission is debris free (clean source of particles rather than the mixture of particles and remnants of solid after laser irradiation) much like as for gases. The high local density of atoms ($n_l \sim 10^{22}$ atoms/cm^3) inside a cluster, lack of thermal dissipation and collective effects (cluster sizes are much smaller than the laser wavelength) are responsible for enhanced energy absorption shown by these clusters. The low average density of atoms ($n_a \sim 10^{15}$ atoms/cm^3) in the clustered beam leads to the debris free emission of particles from cluster explosion. The outcome of the enhanced absorption of incident laser energy is manifested into the emission of highly-charged ions [22] (with energies upto MeV [23]), keV electrons [24] and X-rays ranging from kev [25] to hundreds of eV (<500 eV) [26, 27]. Laser-driven particle accelaration [28], coherent and incoherent X-ray generation [29], nuclear fusion in deuterium clusters

[30], production of efficient plasma wave guides [31] and high orders harmonic generation [32] are few of the important application of laser-cluster interaction.

Most of the early experiments in the field of laser noble gas cluster interaction were carried out with near infrared (NIR) radiation [21, 23, 24, 26, 33–35]. Recently, a new set of experiments are initiated on noble gas clusters with free electron laser such as FLASH at DESY at much shorter wavelengths (VUV, XUV and X-ray) [36–40]. These experiments have raised fundamental questions about wavelength dependence of strong laser induced cluster excitations. These studies are motivated by applications ranging from solid state physics to femtochemistry. The coherent, intense radiation from FEL has the potential to probe the dynamical state of matter and transitions occurring in ultrafast time scale.

In this chapter, we review the basic understanding of noble gas cluster dynamics driven by intense femtosecond laser with the wavelength in NIR domain. We begin by a brief introduction of various fundamental processes occurring in the cluster plasma which is followed by a description of various theoretical models to describe the laser driven cluster dynamics. Next we briefly explain a molecular dynamic (MD) model developed by us. After this, we present our MD results on various interesting features observed in laser driven clusters like anisotropy of ion emission, carrier-envelope phase effect on ionization dynamics of clusters and neutron generation from deuterium clusters.The chapter ends with the summary and conclusion.

4.2 Fundamental Processes Occurring in Laser Driven Clusters

Noble gas clusters are produced by the isentropic expansion of a high pressure gas with sufficiently large polarizability through a nozzle into the vacuum [41, 42]. The decrease of temperature due to the isentropic expansion allows the weak Van der Waals interaction among the gaseous atoms to from clusters. The gas expansion from the nozzle does not produce clusters of identical sizes but they have a long-normal distribution of the form

$$f(N) \propto \exp\left[-\log^2(N/N_0)/2w^2\right], \tag{4.1}$$

where N_0 is the most abundant cluster size and w is proportional to the full width half maximum (FWHM) of the distribution [43]. Hagena and Obert [41, 42] obtained a empirical relationship which characterizes the average size of the cluster. The Hagena parameter is defined as

$$\Gamma^* = k\left[\frac{d(\mu\text{m})}{\tan\alpha}\right]^{0.85}\frac{p_0(\text{mbar})}{T_0^{2.29}(K)}, \tag{4.2}$$

where d is the jet throat diameter, α is the jet expansion half angle, p_0 is the backing pressure, T_0 is the stagnation temperature and k is a constant that depends on the particular atomic species (k = 5500 for Xe, 2890 for Kr, 1650 for Ar, 180 for Ne, and 4 for He [44]). Most studies suggest that clustering begins when this parameter exceeds 300 where the average size of the cluster varies as $(\Gamma^*)^{2.0-2.5}$ [41]. The average size of the cluster in terms of number of atoms per cluster (N_c) is related with Hagena parameter [42, 45] as

$$N_c \cong 33\left[\frac{\Gamma^*}{1000}\right]^{2.35} \qquad (\Gamma^* \leq 10^4) \tag{4.3}$$

$$N_c \cong 100\left[\frac{\Gamma^*}{1000}\right]^{1.8} \qquad (\Gamma^* > 10^4). \tag{4.4}$$

The interaction of laser with clusters proceeds through three processes viz. ionization of atoms, absorption of laser energy, and expansion of cluster. In the ionization process, one has to clearly differentiate between the inner and outer ionized electrons [46]. Inner ionization refers to the removal of electron from parent atom yet confined in the cluster due to coulomb force of residual ions. On the other hand, outer ionization refers to the removal of electron from cluster as the electron achieve sufficient energy to overcome coulomb barrier of the cluster. The initial removal of electron from parent atoms commences through tunnel ionization (TI). The prevalence of TI over MPI is mainly decided by the Keldysh adaibaticity parameter (γ) [47] which is defined as the ratio between the time taken by the electron to tunnel through the barrier ($\tau_{tunnel} = \sqrt{2I_p}/\mathcal{E}_0$) and the laser time cycle,

$$\gamma = \omega\tau_{tunnel}. \tag{4.5}$$

Denoting $U_p = \mathcal{E}_0^2/4\omega^2$ as ponderomotive energy of electron, one can alternatively define γ as

$$\gamma = \sqrt{\frac{I_p}{2U_p}}. \tag{4.6}$$

This parameter defines a transition from MPI ($\gamma \gtrsim 1$) to TI ($\gamma \lesssim 1$). Once sufficient number of electrons are accumulated inside the cluster, they further create collisional ionization [48]. The heating of the cluster plasma is mainly governed by the inner electrons via inverse bremsstrahlung [49] and collective resonance effects [27, 50]. The cluster plasma expansion mode depends upon the population of inner electron. If the space charge is strong enough to retain most of electrons inside the cluster, then cluster plasma expands due to pressure of electron gas. This mode of cluster expansion is known as hydrodynamic expansion. On the other hand, if most of the electrons leave the cluster immediately then cluster explodes due to the repulsion among positively charged ions of the cluster, commonly known as Coulomb explosion. Various reviews [14, 15, 51] provide a overall status of the subject. Below we present a short description of various theoretical models that have been used to explain the various experimental results observed in laser-cluster interaction.

4.2.1 Nanoplasma Model

This model, put forwarded by Ditmire et al. [27], treats the cluster as a small ball of high-density plasma, and so the name nanoplasma. There are two important length scales: one is Debye length of cluster plasma ($\lambda_d = \sqrt{\varepsilon_0 n_e e^2/kT_e} = 5$ Å, for a solid density, 1000 eV plasma) and the other is laser wavelength ($\lambda = 800$ nm). The typical cluster sizes ($\sim$100 Å) lies between the two and hence the assumption holds. The smallness of cluster plasma allows one to ignore the spatial variation of laser field across the cluster and all the atoms and/or ions see the spatially uniform optical field of incident laser. In addition, two assumptions made in this model [27] are (1) high degree of collisionality in the cluster (for the attainment of Maxwellian electrons distribution) and (2) the uniform density within the cluster throughout the cluster expansion. In this model, incident laser deposits its energy to the inner electrons through collisional inverse bremsstrahlung. Assuming the cluster as dielectric sphere, the heating rate averaged over a laser time cycle can be written as

$$\frac{\partial U}{\partial t} = \frac{\omega}{8\pi} \mathrm{Im}\,[\varepsilon]\,[\mathcal{E}]^2 . \tag{4.7}$$

The electric field inside the cluster is then obtained from standard expression for a dielectric sphere in a constant electric field,

$$\mathcal{E} = \frac{3}{|\varepsilon + 2|}\mathcal{E}_0, \tag{4.8}$$

with $\mathcal{E}_0$ as field outside. Substitution of electric field from (4.8) in (4.7) gives the heating rate as

$$\frac{\partial U}{\partial t} = \frac{9\omega}{8\pi} \frac{\mathrm{Im}\,[\varepsilon]}{|\varepsilon + 2|^2} |\mathcal{E}_0|^2 . \tag{4.9}$$

The dielectric constant ε is calculated by the Drude model [52] as

$$\varepsilon = 1 - \frac{\omega_p^2}{\omega\,(\omega + i\nu_{ei})}, \tag{4.10}$$

where ω_p and ν_{ei} represent the plasma frequency and electron-ion collision frequency. It is instructive to see that both electric field and heating rate goes to a maximum when $|\varepsilon + 2|$ goes to a minimum, which is a clear indication of resonance. The minimum value of $|\varepsilon + 2|$ corresponds to plasma frequency $\omega_p = \sqrt{3}\omega$ or in terms of electron density $n_e = 3n_c$, where ω and n_c represent the laser frequency and critical density, respectively. Initially the cluster plasma is overcritical i.e. $n_e \gg 3n_c$ but as the plasma expands the electron density reduces and resonance occurs at $n_e = 3n_c$.

The condition of resonance can also be obtained by microscopic modelling of the motion of inner electron cloud along the laser polarization direction as a driven and damped harmonic oscillator [50]. The eigenfrequency ($\omega_t = \sqrt{Q_{ion}/R(t)^3} = \omega_p/\sqrt{3}$) of this oscillation is determined by total ionic charge (Q_{ion}) and size (R) of the cluster. The condition of resonance is achieved when this eigenfrequency (ω_t) matches with the laser frequency (ω) due to the electron emission and cluster expansion.

The cluster expands under the combined effect of hydrodynamic pressure ($P_e = n_e k T_e$) of heated electron gas and Coulomb pressure ($P_{Coul} = Q^2 e^2/8\pi r^4$) from a charge[2] build-up on the cluster because of outer ionization. For small clusters Coulomb force dominates ($1/r^4$ scaling law). However, the hydrodynamic force will begin to dominate as cluster expands (hydrodynamic pressure scales as $1/r^3$).

4.2.2 *Modifications to Nanoplasma Model*

One of the main implications of the uniform density nanoplasma model is the prediction of a narrow resonance time interval, [53]

$$\delta t_{res} = \frac{2}{3}\frac{\nu_{ei}}{\omega}\left(\frac{n_e}{3n_c}\right)^{1/3}\frac{R}{c_s} \tag{4.11}$$

for a cluster of radius R and electron density n_e, where c_s is the plasma sound speed, and ν_{ei}/ω is the normalized electron-ion collision frequency at $n_e = 3n_c$. Model calculations suggest that this time interval (δt_{res}) is 6 and 40 fs for clusters of radii 100 and 600 Å, respectively. The pump-probe and variable pulse width experiments suggested a much longer time interval (severals of picoseconds) for these resonances [35]. This discrepancy was subsequently resolved by including the radial non-uniformity of cluster expansion in self-consistent one dimensional model [53, 54]. These studies showed that even for smaller clusters, radial non-uniformity of electron density is important and its inclusion results into *long-time resonances* at the critical density plasma layer n_c in stead of $3n_c$. This resonance is maintained throughout the pulse duration as long as the cluster plasma does not expand below critical density.

The nanoplasma model employs quasi-static dielectric constant given by Drude's model ($\varepsilon = 1 - \omega_p^2/\omega(\omega + i\nu_{ei})$) which overestimates the enhancement of electric field in clusters. Instead of this, Liu et al. [55] modified the Ditmire's nanoplasma model by inclusion of an effective plasma dielectric constant ($\varepsilon_{eff} = \varepsilon + i\varepsilon/\omega$), which

[2]The total charge (Q) build up on the cluster is sum of total ionic charge (Q_{ion}) and the charge of inner electrons.

is obtained by solving the Maxwell's equations for a rapidly expanding plasma. It can be shown that the electric field calculated by using ε_{eff} is much smaller than that by using ε.

Megi et al. [56] suggested that the consideration of other damping mechanism like electron cluster surface collision other than the usual electron ion collision frequency ν_{ei} to explain the puzzling issue of electron heating in the domain of ν_{ei} approaching to zero. In fact, the damping effect can be accounted for by an effective collision frequency, $\nu_{eff} = \nu_{ei} + Av/R$, where v is the effective velocity of electron (vector sum of electron thermal velocity and electron quiver velocity) and A is a model parameter close to unity. The second term in ν_{eff} signifies the contribution from collisions with cluster surface. ν_{ei} reduces near the resonance due to increase in both the laser intensity and electron quiver velocity but the surface term increases with the electron velocity. Thus the total collision frequency does not drop significantly near the resonance which prevents the electric field to diverge inside the cluster.

In addition of first-order ionization processes included in the original framework of nanoplasma model, the ionization via intermediate excited states was suggested by Micheau et al. [57] to facilitate the high-order ionization observed in the experiments. The effect of lowering of ionization potential in the dense cluster environment was studied by Hilse et al. using a modified nanoplasma model [58]. In the regime of the dense non-ideal nanoplasma, the bound state properties can be strongly affected by the interaction with the surrounding particles. An important effect is the ionization energy suppression which was treated by Hilse et al. using the Stewart-Pyatt approach [59]. The main result was the appearance of considerable higher ionic charge states if non-ideal rates[3] are used in the model.

4.2.3 Non-linear Models

Mulser et al. [60] suggested that the nanoplasma model based upon linear resonance [27, 50] does not work for ultra-short laser pulse duration or during the early-cycles of a long-pulse laser cluster interaction due to the condition $n_e \gg n_c$ or equivalently $\omega_t \gg \omega$. To account for the laser energy absorption in this regime, idea of non-linear resonance (NLR) [60] i.e. lowering of the eigenfrequency or resonance frequency with increasing oscillation amplitude was proposed. To explain NLR, a rigid sphere model of two inter-penetrating spheres (RSM) of positive charge (ions) and sphere of negative charge (electron) was considered [61]. The potential between two spheres for small energy of excitation is harmonic with eigenfrequency $\omega_t \gg \omega$ but it becomes anharmonic for higher oscillation amplitude, leading to a weaker restoring force. Consequently, the eigenfrequency decreases and it may become equal to laser frequency at some excitation energy leading to the NLR.

[3] Non-ideal rates are electron impact ionization rates that incorporate the shift of ionization potential in the dense plasma system.

4.2.4 Particle-Particle/Particle-Mesh Models

Nanoplasma model and its different variants, being fluid models, can only describe successfully an average picture of laser-cluster dynamics. Particle models have been developed to get microscopic details of the interaction which have been categorised in two types: particle-particle (e.g. molecular dynamics) [46, 50, 62–66] and particle-mesh (e.g. particle-in-cell) [67–70].

Once inner ionization of cluster is complete, both molecular dynamics (MD) and particle-in-cell (PIC) use classical equations of motion to advance the particle in time domain. The difference lies in the calculation of total force appearing in the equation of motion. In MD, the force on each particle is the sum of all two-body Coulomb forces from other charged particles which is $O(N^2)$ for N number of particles in the cluster. The force due to the laser electromagnetic field is also added to this force. In the PIC, the field is computed on a grid of cells, and the force on the particles is calculated from the interpolated fields at the particle position. The computation time is $O(N)$ for N number of particles which clearly establishes better efficiency of PIC over MD for large number of particles. On the other hand, MD methods are grid-less calculations, facilitating their scalability to three dimensions much more easily compared to PIC. MD methods are favoured over PIC for limited number of particles (10^3–10^5). It is noteworthy to mention that N^2 problem of MD calculations can be improved by using Barnes and Hut algorithm [71] in which the computation time is $O(N \log N)$.

4.3 Molecular Dynamic Simulations

4.3.1 Basic Model

In our MD model [72, 73], we consider a single rare gas spherical cluster of radius R_0 which is irradiated by a high intensity laser pulse propagating along y-direction and polarized in x-direction. The center of the cluster is assumed to be at the origin of the coordinate system ($x = y = z = 0$). The number of atoms in a cluster is calculated by the knowledge of Wigner-Seitz radius R_W as, $N_{Atoms} = (R_0/R_W)^3$. The value of R_W depends on the species of the gas cluster under study. For example R_W is 1.70, 2.02, 2.40 and 2.73 Å for Deuterium, Neon, Argon and Xenon clusters respectively [74]. For computational purpose we use predefined number of macro particles, N_{Macro}. Each macro particle consists of N_{Atoms}/N_{Macro} of actual particles (ions or electrons). This procedure of lumping of actual particles to form pseudo-particles is similar to the one presented by Last and Jortner [75]. The presence of other clusters in the neighbourhood are simulated by imposing a periodic boundary conditions on the faces of cubical simulation box with each side equal to the inter cluster distance, R_{IC}. The temporal profile of incident laser pulse is assumed to be Gaussian. The spatial variation of laser intensity is ignored as the size of the cluster is much smaller than the laser wavelength.

The inner ionization of the cluster atoms is carried out by tunnel and collisional ionization using monte-carlo method. For the tunnel ionization, we calculate the ionization rate as a function of laser field strength by using ADK tunnel ionization formula [76],

$$\nu_{ofi} = \left(\frac{3\mathcal{E}_0 n^{*3}}{\pi Z^3}\right)^{1/2} \frac{(2l+1)(l+|m|)!}{2^{|m|}(|m|)!(l-|m|)!} \left(\frac{2e}{n^*}\right)^{2n^*} \frac{1}{2\pi n^*} \left(\frac{2Z^3}{\mathcal{E}_0 n^{*3}}\right)^{2n^*-|m|-1}$$
$$\times \left(\frac{Z^2}{2n^{*2}}\right) \exp\left(-\frac{2Z^3}{3n^{*3}\mathcal{E}_0}\right) \frac{1}{2.4\times 10^{-17}} \quad s^{-1}, \tag{4.12}$$

where $\mathcal{E}_0 = \sqrt{2\times 10^4\ I/(c\ \varepsilon_o)}/5.14\times 10^{11}$ is the electric field in atomic units, I (W/cm^2) is the intensity of laser light and c is the speed of light in SI units, ε_o is the permittivity of free space, Z is the degree of ionization, $n^* = Z/\sqrt{2I_p}$ is the effective quantum number, I_p is the ionization potential in atomic units, e is the charge of electron, ℓ and m are the orbital and magnetic quantum numbers of outer electron respectively. Ionization rate is calculated for each value of m and averaged over it. The collisional ionization rate is calculated from the best fitted data on the ionization rates [77] for the atomic species of the cluster whose $Z < 28$. For high Z clusters, Lotz formula for collisional ionization rate is used. It is given as [48],

$$\nu_{ci} = 6.7\times 10^{-7} \frac{n_e\, a\, q}{I_p\sqrt{kT_e}} \int\limits_{I_p/kT_e}^{\infty} \frac{e^{-x}}{x} dx \quad s^{-1}, \tag{4.13}$$

where n_e (cm^{-3}) is electron density, $a = 4.5$ is constant, q are number of electrons in outer orbital of particular ion, I_p (eV) is the ionization potential for a particular ion, kT_e (eV) is the electron energy. Particles inside the cluster experience both the electromagnetic force due to laser and electrostatic Coulomb force due to presence of other charged particles inside it. The charged particles move according to their relativistic equation of motion ($d\mathbf{p_i}/dt = \mathbf{F_i}$, $\mathbf{v_i} = \mathbf{p_i}/(m_i\sqrt{1+|\mathbf{p_i}|^2/(m_i c)^2})$, $d\mathbf{r_i}/dt = \mathbf{v_i}$; $\mathbf{p_i}$, $\mathbf{v_i}$, $\mathbf{r_i}$ and m_i are relativistic momentum, velocity, coordinate and mass of the ith particle) and the phase-space of all the particles is stored after certain time-steps.

4.3.2 Anisotropic Ion Emission

The mode of the cluster expansion, both hydrodynamic as well as Coulomb, leads to the isotropic emission of ions (Fig. 4.3a). But there are many instances where the number of inner electrons is less than that required for pure hydrodynamic expansion and more than that required for the pure Coulomb explosion. In this situation, the cluster ions show certain preferred direction for emission. This is known as anisotropic cluster explosion and it depends strongly on the pulse duration of laser.

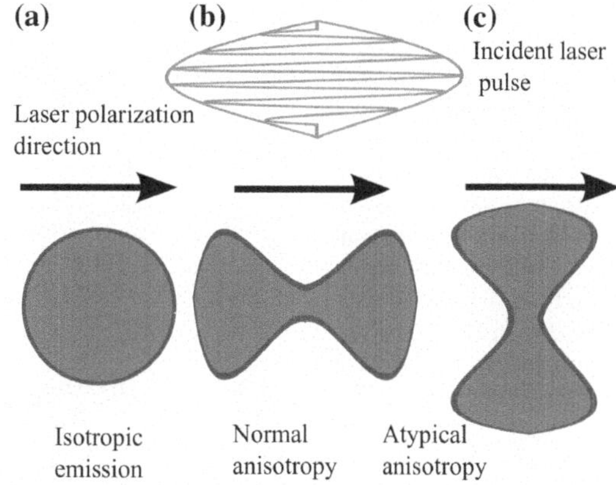

Fig. 4.3 Schematic showing the difference among isotropic emission, normal anisotropy and atypical anisotropy

For many cycles pulse duration, ion yield was observed to be more along the direction of laser polarization (0°) than perpendicular (90°) to it [78–81] and it is termed as "Normal anisotropy" (Fig. 4.3b). When the laser pulse duration approaches to few cycles, the nature of the anisotropy reverses i.e. more number of ions are observed along the perpendicular direction of laser polarization than parallel to it [82–84] and it is termed as "Atypical anisotropy" (Fig. 4.3c).

4.3.2.1 Normal Anisotropy

In an experiment performed by Kumarappan et al. [79], the problem of "Normal anisotropy" was studied for Ar clusters (N $\sim 2 \times 10^3$ to 4×10^4) irradiated by laser of intensity 8×10^{15} W/cm^2 and pulse duration 100 fs. Experimental observations revealed that the ion energy distribution consisted of two components: a low energy isotropic component, and a high energy anisotropic one. To explain the anisotropy results, Kumarappan et al. used two earlier theoretical results of Ishikawa et al. [63] and Kou et al. [85].

The net field (laser field plus radial Coulomb field of charged particles) experienced by the ions is different in two consecutive laser half time cycles along the laser polarization direction and this causes a rapid change in the charge state of ions in resonance with the optical field. Consequently, the ions feel a net cycle averaged force due to the laser field along the laser polarization direction. In addition, the charge state distribution itself is asymmetric i.e. charge state are higher along the polarization direction. Consequently, the Coulomb explosion is asymmetric [85]. Taken together these two effects, Kumarappan et al. were able to explain the anisotropy for the high energy ions that ejected from the surface of the cluster. For the ions interior to the cluster, the laser field is sufficiently shielded by the electrons and only the radial field remains into the picture. Consequently, the inner part of the cluster giving rise to low energy ions show less anisotropy than that depicted by high energy ions emitting from the surface of the cluster.

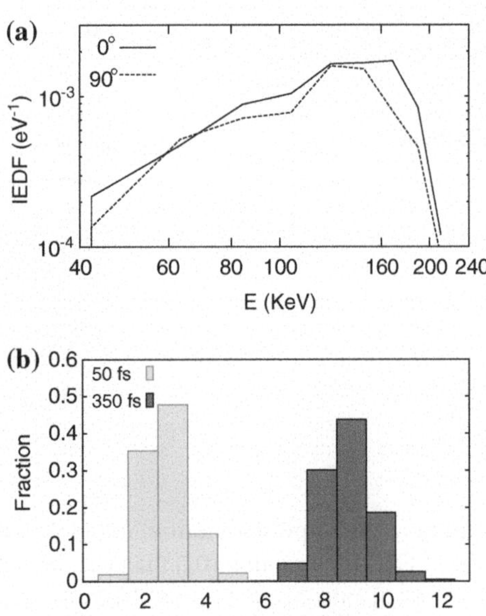

Fig. 4.4 The IEDF (**a**) for 82 Å Argon cluster along (0°) and perpendicular (90°) to laser polarization direction and the fractional charge states at 50 and 350 fs (**b**) (Reprinted with permission from (Amol R. Holkundkar, Gaurav Mishra, N.K. Gupta, Phys. Plasmas, Vol. 18, Page 053102-1-7 (2011)). Copyright (2011), American Institute of Physics)

We have used our MD code to study this kind of anisotropy and the corresponding ion energy distribution function (IEDF) along 0° and 90° is shown in Fig. 4.4a while the fraction of different charge states at 50 and 350 fs are shown in Fig. 4.4b. Comparison of our MD results of IEDF with experimental IEDF of Kumarappan et al. show striking similarities. Our MD results correctly reproduces the experimental ion energies of 200 keV. Moreover, the observed asymmetry along the laser polarization direction which was the essence of the experiment [79], is also verified in our results. As can be seen from Fig. 4.4b, in early stages most of the ions are doubly and triply ionized which in due course of time are enhanced up to 10. The charge states so obtained are also in good agreement with the PIC study of same laser and cluster conditions [86].

The anisotropic studies on hydrodynamic expansion of Xe clusters were also performed by Kumarappan et al. [80]. They extended the hydrodynamic model [27] of cluster expansion and found the importance of the polarization induced surface charge distribution. The action of incident laser field on this induced surface charge distribution gives rise to a additional directional dependent cycles averaged pressure ($\propto \cos^2\theta$). This additional pressure has to be added to already present hydrodynamic pressure to account for the observed anisotropy.

4.3.2.2 Atypical Anisotropy

Recent experimental results [82, 83] on Ar_n clusters ($n = 400$–900) and Xe_n clusters ($n = 500$–$25{,}000$) driven by 800 nm laser at peak intensity of $\approx 5 \times 10^{14}$ W/cm^2

with pulse duration in few cycles limit (∼3–4 laser time cycles) demonstrate the complete reversal of the anisotropy of ion emission that was observed earlier with comparatively longer pulse duration (∼100 fs). They observed that ion yields were larger when the polarization of the laser is perpendicular to the detection axis than along it. This "Atypical anisotropy" was explained qualitatively on the basis of spatially anisotropic shielding of ions due to the oscillating electron charge cloud within the cluster [82, 87]. Similar studies were also performed by Skopalova et al. [84] with Xe and Ar clusters. In these experiments, they observed that ion emission was more energetic in the direction perpendicular to the laser polarization than parallel to it. They also interpreted these result phenomenologically in terms of screening of the ionic field along laser polarization direction due to the collective oscillation of electron cloud.

We have performed anisotropic studies on Ar [87] and Xe [88] clusters driven by laser pulses in few cycles time duration regime. Let us discuss the results of anisotropic ion emission from 56 Å Xe cluster exposed to laser pulse of intensity 4.5 $\times$ 10^{15} W/cm^2 [88]. The pulse duration of the incident laser beam is varied from 10 to 100 fs. The temporal evolution of integrated yield of ions for the above mentioned pulse durations are shown in Fig. 4.5a–d. We note that the ion yield for shortest pulse duration (τ = 10 fs) remains higher along the direction perpendicular to laser polarization than parallel to it while opposite is observed for longest pulse duration (τ = 100 fs). These results can be explained on the basis of spatial shielding of ions by collective oscillations of inner electrons in the presence of external laser electric field. For few cycle laser pulses, incident laser field terminates before any significant outer ionization. This in turn leads to build up of inner electrons. This inner electron cloud oscillates in harmony with the external laser electric field. Collision effects of electrons with other particles is also not significant for the few cycle laser pulse duration that can otherwise hampers the directed motion of electron cloud. Consequently, the inner electron charge elongates itself along the direction of laser electric field due to their continuous rapid oscillation in harmony with the laser field. The smearing of the electronic charges shields the ions effectively along laser polarization than perpendicular direction. This initial shielding of ions results into the observed anisotropy of ions for the few cycle laser pulse. As the pulse duration is increased, collision effects start affecting the directional motion of electron cloud that results into the reduced shielding of ions due to the oscillating electron cloud. This effect reduces the observed preferential direction of cluster explosion. For still longer pulse durations, surface effects start dominating the ion anisotropy [80] as explained previously.

To understand this shielding phenomenon in more detail, the knowledge of electric field at the periphery of expanding cluster will be more useful. Therefore, we calculate the time variation of electric fields along the direction parallel to laser polarization (E_0) and perpendicular to it (E_{90}) for laser pulse durations of 10 and 100 fs (Fig. 4.6). We observe that E_{90} is more than E_0 for pulse duration of 10 fs (Fig. 4.6a) whereas these two fields do not differ significantly for pulse duration of 100 fs (Fig. 4.6b). These variations of electric fields explain the anisotropy in the ion emission for pulse durations of 10 and 100 fs.

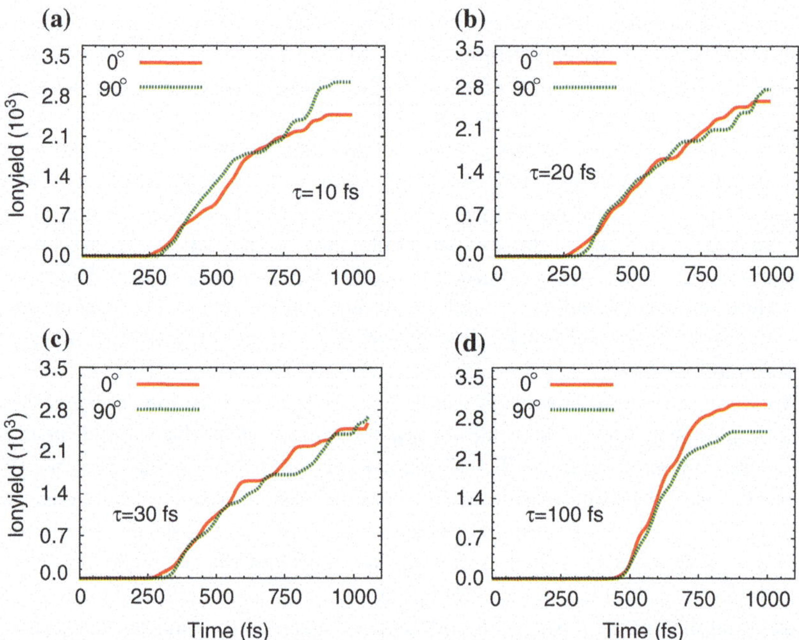

Fig. 4.5 Temporal evolution of integrated ion yields for 56 Å Xe cluster exposed to laser of intensity 4.5×10^{15} W/cm^2 for FWHM pulse duration of 10 fs (**a**), 20 fs (**b**), 30 fs (**c**) and 100 fs (**d**) (Reprinted with permission from (Gaurav Mishra and N.K. Gupta, Phys. Plasmas 19, 093107 (2012)). Copyright (2012), American Institute of Physics)

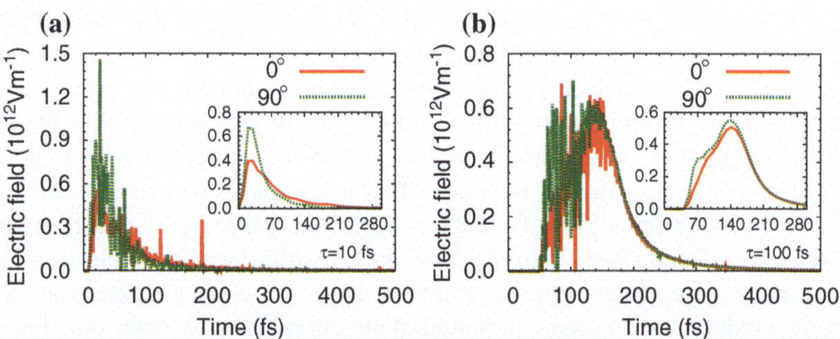

Fig. 4.6 Temporal variation of electric field at the surface of the cluster for laser pulse durations of 10 fs (**a**), and 100 fs (**b**). The other laser and cluster parameters are same as used in Fig. 4.5. In the *inset* of each plot, we have shown the smoothed electric field till 280 fs (Reprinted with permission from (Gaurav Mishra and N.K. Gupta, Phys. Plasmas 19, 093107 (2012)). Copyright (2012), American Institute of Physics)

4.3.3 Carrier-Envelope Phase Effects

Presently, it is possible to generate the laser pulses with time duration up to few laser time cycles [89]. For such ultrashort laser pulse, the complete description of laser electric field ($\mathcal{E}(t) = \mathcal{E}_0(t)\cos(\omega t + \phi)$) requires the knowledge of initial phase (ϕ) of the carrier with respect to the envelope, commonly known as CE phase, in addition to field amplitude ($\mathcal{E}_0$) and carrier frequency (ω). Thus the strong field interactions originating from few cycles laser pulses become sensitive to the correct value of the CE phase. Generally, it is not possible to generate the laser pulses in succession with fixed initial CE phase from the oscillator due to the difference between the phase velocity and group velocity. Due to this, the carrier wave slides under the envelope as the pulse circulates in the mode-locked laser [90] and pulse to pulse CE phase changes accordingly. Now, it has been possible to control this parameter by appropriate feed back techniques [91, 92] and also, direct measurement of ϕ has been demonstrated [93, 94].

When these CE phase stabilised laser pulses are coupled with the individual atoms, it lead to the very interesting phenomena. For such ultra-short laser pulses with stabilised laser pulses, the probability of tunnelling of the electron from the parent atoms depends on the value of the instantaneous electric field. Thus the release time of the electron from the parent atoms is dependent on the instantaneous electric field that inherently contains the information of ϕ. This released electron while following the laser field, may return to the parent atom depending on the time of its ejection within the laser cycle. When the electron approaches to the atom, it can lead to high harmonic generation (HHG) by recombination, above-threshold ionization (ATI) by elastic scattering or non-sequential double ionization (NSDI) by inelastic scattering with it [95]. It has been shown experimentally that all of these processes strongly depend on the CE phase [93, 94, 96].

Instead of isolated atoms, we have used small Xe clusters in our MD studies and tried to see the effect of CE phase on the ionization dynamics of these clusters [97]. We have simulated the ionization dynamics of Xe_{400} clusters driven by laser pulse of few cycles ($\tau_1 = 2T_0$, where $T_0 = 2.67$ fs is one laser time cycle for $\lambda = 800$ nm) and many cycles ($\tau_2 = 8T_0$). For numerical convenience, we have used $\sin^2$ pulse instead of Gaussian laser pulse. The temporal variations of instantaneous electric fields along with their envelopes are shown in Fig. 4.7a, b for $\tau = 2T_0$ and $\tau = 8T_0$, respectively. For each pulse duration, these plots also show field variations for CE phases of $\phi = 0$ and $\phi = \pi/2$ separately. The traditional ADK TI rate (4.12) can not be used for few cycles laser pulses as these are the cycle averaged rates determined in the adiabatic approximation[4] which holds well for much longer laser pulse durations. The electric field appearing in the TI rate is amplitude of the electric field which does not contain the CE phase information. For the case of ultrashort laser pulses, the validity of adiabatic approximation is questionable. As a first approximation, we use absolute value of time dependent electric field ($|E(t)|$) [98] in ADK rate (4.12). It is important

[4]Adiabatic approximation is valid when tunnelling time is much smaller than laser time cycle so that electron sees a constant electric field at the time of tunnelling.

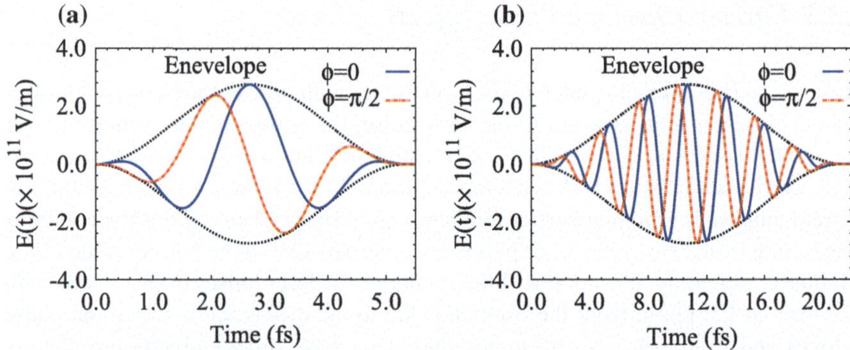

Fig. 4.7 Time envelope and instantaneous electric field for laser of pulse durations $2T_0$ (**a**) and $8T_0$ (**b**). The peak intensity of the laser pulse is kept same at 1×10^{16} W/cm^2 for both the cases. *Solid blue* and *red dot dashed lines* represent the electric field variations for CE phases of $\phi = 0$ and $\phi = \pi/2$, respectively. The electric field envelope is shown by the *dotted line* (Gaurav Mishra and N.K. Gupta, J. Phys. B: At. Mol. Opt. Phys., 46, 125602 (4 June 2013) ©IOP Publishing. Reproduced by permission of IOP Publishing. All rights reserved)

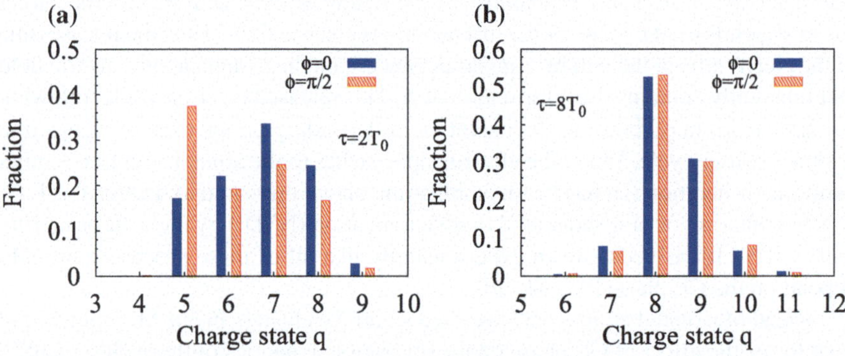

Fig. 4.8 Fraction of various charge states for Xe_{400} cluster irradiated by laser of pulse durations $2T_0$ (**a**) and $8T_0$ (**b**). *Solid blue* and hatched *red bar* correspond to CE phases of $\phi = 0$ and $\phi = \pi/2$, respectively. The intensity of the laser is same as used in Fig. 4.7. Absolute value of time dependent electric field ($|E(t)|$) is used in ADK tunnel ionization rate to account for the effect of the CE phase (ϕ) on ionization dynamics of Xe_{400} cluster (Gaurav Mishra and N.K. Gupta, J. Phys. B: At. Mol. Opt. Phys., 46, 125602 (4 June 2013) ©IOP Publishing. Reproduced by permission of IOP Publishing. All rights reserved)

to note that collisional ionization also becomes indirect function of CE phase as the energy of the colliding electron depends upon the birth in the laser cycle. We show the results of ionization dynamics for the two cases of pulse durations with different CE phases. The results are shown in Fig. 4.8a, b. We clearly observe a significant difference in fractional ionization of various charge states for the two values of CE phase when the few cycles laser pulse ($\tau_1 = 2T_0$) is used for irradiation. For the case of many cycles ($\tau_2 = 8T_0$), these effects average out so that we do not observe any noticeable difference in fractional ionization.

4.3.4 Neutron Generation from Deuterium Clusters

As mentioned in the Sect. 4.1, nearly 100 % absorption of the incident laser energy [21] leads to the generation of highly energetic ions in the range of 10 keV to 1 MeV for large Xe clusters [23]. In the case of explosion of deuterium and/or tritium clusters, the multi keV ions can lead to generation of neutrons via the various nuclear fusion reactions ($D + D \xrightarrow{50\,\%} He^3 + n$, $D + D \xrightarrow{50\,\%} T + p$, $D + T \rightarrow He^4 + n$). The use of laser driven deuterium clusters as a neutron source is motivated by various factors like table-top dimensions of the whole set-up, nearly point like emission, almost monochromatic energy distribution, high-repetition rate and temporal durations as short as a few hundred picoseconds [99]. Such a short, nearly point like neutron source offers its use as a source of neutron pulse to carry out neutron damage pump probe experiments [99]. It also offers a wide range of applications in material science and neutron radiography [30].

The first experimental demonstration of laser irradiated clusters using as neutron sources (Fig. 4.9) was shown by pioneering experiment performed by Ditmire and his group [30]. The average size of the deuterium clusters used in this experiment was nearly 50 Å while the peak intensity of the laser was about $2 \times 10^{16}\,W/cm^2$. The Coulomb explosion of clusters present in the focal volume of the laser leads to the formation of a cylindrical plasma filament and highly energetic ions from the neighbouring clusters in this filament fuse together to give neutrons with characteristics energy of 2.45 MeV. Experiments by Zweiback et al. [100] established a strong dependence of neutron yield on the cluster size which was found to be consistent with the Coulomb explosion model. According to this model, the energy of deuterons increases with cluster size and fusion cross section is a strong function of ion kinetic energy. Consequently, the neutron yield becomes a sensitive function

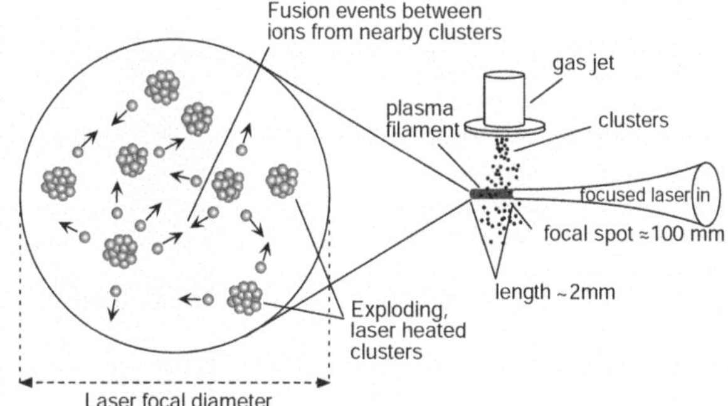

Fig. 4.9 Layout (Reprinted by permission from Macmillan Publishers Ltd: (Nature) (T. Ditmire, et al., 398, 489 (1999), copyright (1999))) of the deuterium cluster fusion experiment conducted by Ditmire et al. [30]

of cluster radius. Madison et al. investigated the role of laser pulse duration on the neutron yield [101] and found, both the cluster explosion energy and resultant fusion yield, to be dependent on the laser pulse rise time: a parameter related to the pulse duration of the laser. In the limit of laser pulse rise time approaching towards zero, both the cluster ionization time and its intrinsic expansion time gets decoupled and this leads to maximum Coulomb explosion kinetic energies of ions. In any real situation, these two time scales start overlapping and the final kinetic energies of the ions after Coulomb explosion would always be smaller than the maximum. This leads to reduced neutron yield for longer pulse duration. Molecular dynamic studies performed by Last and Jortner [102] found enhanced neutron yield from the Coulomb explosion of heteronuclear clusters $(D_2O)_n$, as compared with homonuclear clusters $(D)_n$. The increased neutron yield for $(D_2O)_n$ clusters was the result of higher kinetic energies of D^+ ions triggered by the highly charged O^{+q} ions. By using molecular dynamic model, Petrov and Davis [103] studied the neutron production from high intensity laser-cluster interaction in an alternate beam-target geometry (Fig. 4.10). They used the laser driven clusters as a source of high energy deuterium ions which reacted with the walls of a surrounding fusion reaction chamber with walls coated with DT fuel or other deuterated material such CD_2 and generated a large amount of neutrons.

We have investigated the neutron generation by deuterium clusters in beam target geometry as suggested by Petrov and Davis [103]. The existing MD code [72, 73] is coupled with a one dimensional beam energy deposition model [104]. In this model, the total number of neutrons produced per pulse in a fusion reaction is given as, $Y = N_d\langle y\rangle$, where N_d is the number of the deuterons produced from the exploding cluster and $\langle y\rangle$ is the average reaction probability. The average reaction probability can also be thought as the neutron yield per ion and it is defined as, $\langle y\rangle = \int_0^{E_{\max}} P(E)y(E)dE$, where $P(E)$ denotes the energy distribution of the cluster ions with maximum energy as $E_{\max}$. $y(E)$ appearing in this equation represents the reaction probability for an ion with the initial energy E penetrating into target, $y(E) = \int_0^E \sigma(E')N_0/S(E')dE'$, where N_0 is the target density, $\sigma(E)$ is the fusion cross section of the deuterium ions

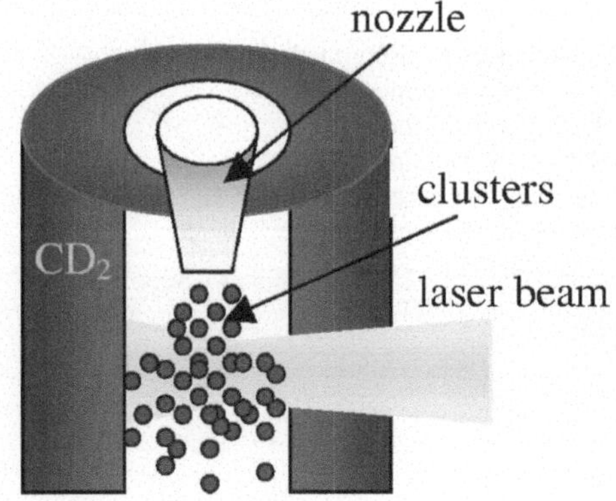

Fig. 4.10 Schematic (Reprinted with permission from (J. Davis, G.M. Petrov, and A.L. Velikovich, Phys. Plasmas 13, 064501 (2006)). Copyright (2006), American Institute of Physics) of the proposed geometry by Davis et al. [103] for neutrons generated from laser driven cluster

with target material and $S(E)$ is the stopping power of the target material. The fusion cross-section σ is calculated by the fitted expression by Huba [105]. The stopping power of deuterium ions in the target material of density 5×10^{22} cm^{-3} is calculated by SRIM code [106]. The total number of deuterons required for the calculation of neutron yield is estimated as $N_d = \eta E_{Laser}/E_{ave}$, where E_{Laser} and E_{ave} denote the energy of the laser pulse and average kinetic energy of the ions respectively. η appearing in this equation represents the fraction of the absorbed energy which is transferred to ions.

We have investigated the dependence of neutron yield on various laser and cluster parameters like laser intensity, cluster radius and cluster density with this modified version of MD code. The former two parameters are straight forward while the effect of cluster density is studied by varying the inter-cluster distance which is same as the size of the simulation box used in the MD model. In particular, the effect of average cluster density ($N_c ls$) is investigated by simulating the interaction dynamics of 100 Å deuterium cluster driven with laser of peak intensity 10^{18} W/cm^2 and pulse duration 50 fs for various values of R_{IC} in multiple of cluster radius. The corresponding results of average kinetic energy of ions and total laser energy absorbed by the cluster are presented in Fig. 4.11. We observe that the average kinetic energy varies as $\sim R_{IC}^3$ that can be explained on the basis of the shielding effect from neighbouring cluster electrons. The smaller is the inter-cluster distance, the larger is the number of clusters per unit volume with same interaction dynamics. The coulomb field responsible for the cluster explosion will be shielded by the electrons of the neighbouring clusters and this results in the poor average kinetic energy of the cluster ions. As R_{IC} increases, the clusters become more sparse in the interaction region and this reduces the shielding of ions by electrons of neighbours. Consequently, the average kinetic energy of the ions

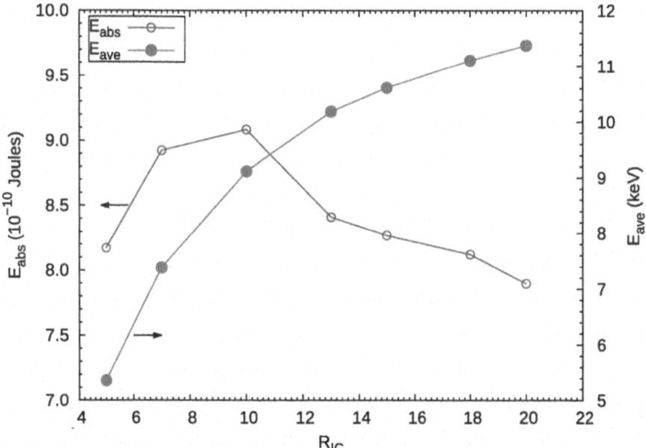

Fig. 4.11 Variation of absorbed laser energy and average kinetic energy with inter-cluster distance ($R_{IC} = n \times R_0$, n is an integer). The size of the cluster is (R_0) is 100 Å whereas the intensity and pulse duration of the laser are chosen as 10^{18} W/cm^2 and 50 fs (Reprinted with permission from (A.R. Holkundkar, Gaurav Mishra and N.K. Gupta, Phys. Plasmas 21, 013101 (2014)). Copyright (2014), American Institute of Physics)

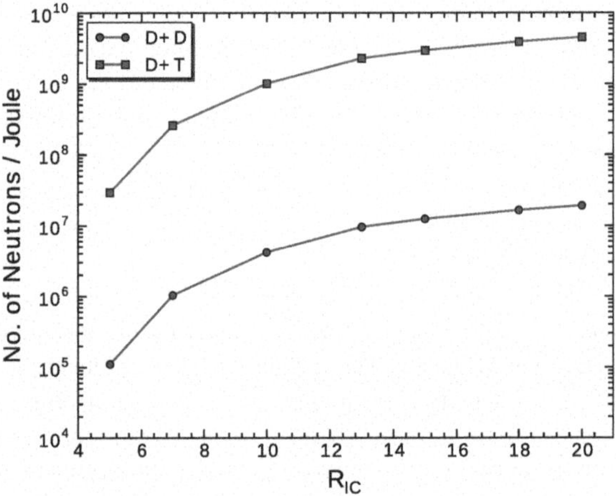

Fig. 4.12 Variation of neutron yield (Reprinted with permission from (A.R. Holkundkar, Gaurav Mishra and N.K. Gupta, Phys. Plasmas 21, 013101 (2014)). Copyright (2014), American Institute of Physics) with inter-cluster distance for laser and clusters parameters as used in Fig. 4.11

increases. However as can be seen in Fig. 4.11, the absorption is not monotonically changing as we vary the inter-cluster distance and there is an optimum inter-cluster distance at which we see a relatively larger absorption. This optimum inter-cluster distance can be understood in terms of the interplay between kinetic and potential energy of the system. For lower R_{IC} the main contribution in absorbed energy is because of the kinetic energy of the particles, however for larger R_{IC} the drop in potential energy will effectively lower the amount of energy absorbed. The neutron yield for various values of the inter-cluster distance is presented in Fig. 4.12. It is observed that for higher number of clusters per unit volume (lower R_{IC}) the neutron yield is small as compared to the low cluster densities. It is mainly because the average kinetic energy is low for the case of low R_{IC} as observed in Fig. 4.11.

4.4 Summary and Conclusions

Intense laser pulse interaction with rare-gas clusters is a challenging subject involving the nonlinear, nonperturbative response of many ions and electrons. We described various theoretical models to explain the experimental features observed in laser-cluster interaction. We have also presented a brief description of MD model developed by us for this purpose. Using the MD model, we were able to observe the normal anisotropy (ion yield more along parallel to laser polarization than perpendicular to it) in laser pulse duration with hundreds of femtosecond. We have explained the atypical anisotropy (ion yield more along perpendicular to laser polarization than perpendicular to it) observed in the few cycles pulse duration regime. The oscillating inner electrons along the laser polarization direction shield the ion along this direction that leads to more emission of ions along perpendicular to laser polarization direction. We also investigated the effect of carrier-envelope phase on the ionization dynamics of Xe cluster in many cycles and few cycles pulse duration regime. The fraction of various ionic species was observed to be different when the CE phase was changed from 0 to $\pi/2$ in the few cycle regime. No difference was observed for many cycles pulse duration. We also studied the neutron generation from deuterium clusters and evaluated the effect of various parameters like cluster size, laser intensity, cluster density on the neutron yield. In particular, neutron yield increases as the cluster density reduces or inter-cluster increases which is explained on the basis of reduced shielding of ions due to the electrons of neighbouring clusters.

Acknowledgments We acknowledge many useful discussions and suggestions from Dr. Amol R Holkundkar. We are thankful to Dr. G.M. Petrov for useful inputs at the time of code development. We also acknowledge Prof. Deepak Mathur for suggesting the problem of anisotropic ion emission from clusters.

References

1. D. Strickland, G. Mourou, Opt. Commun. **56**, 219 (1985)
2. P. Maine et al., IEEE J. Quantum Electron. **24**, 398 (1988)
3. D.E. Spence et al., Opt. Lett. **16**, 42 (1991)
4. M. Protopapas et al., Rep. Prog. Phys. **60**, 389 (1997)
5. P. Agostini et al., Phys. Rev. Lett. **42**, 1127 (1979)
6. A. McPherson et al., J. Opt. Soc. Am. B **4**, 595 (1987)
7. M. Ferray et al., J. Phys. B At. Mol. Opt. Phys. **21**, L31 (1988)
8. A.D. Bandrauk, *Molecules in Laser Fields* (5. Marcel Dekker Inc., New York, 1994)
9. W. Kruer, *The Physics of Laser Plasma Interactions* (Reading, MA (US), Addison-Wesley Publishing Co., 1988)
10. M.M. Murnane et al., Science **251**, 531 (1991)
11. B.M. Hegelich et al., Nature **439**, 441 (2006)
12. V. Malka et al., Nature **431**, 541 (2004)
13. B.A. Remington et al., Science **284**, 1488 (1999)
14. V.P. Krainov, M.B. Smirnov, Phys. Rep. **370**, 237 (2002)
15. T. Fennel et al., Rev. Mod. Phys. **82**, 1793 (2010)
16. R.L. Johnston, *Atomic and Molecular Clusters* (Taylor & Francis, New York, 2002)
17. F. Calvayrac et al., Phys. Rep. **337**, 493 (2000)
18. P.-G. Reinhard, E. Suraud, *Introduction to Cluster Dynamics* (WILEY-VCH Verlag GmbH & Co. KGaA, Weinheim, 2004)
19. L. Köller et al., Phys. Rev. Lett. **82**, 3783 (1999)
20. T. Döppner et al., Phys. Rev. Lett. **94**, 013401 (2005)
21. T. Ditmire et al., Phys. Rev. Lett. **78**, 3121 (1997)
22. T. Ditmire et al., Phys. Rev. Lett. **78**, 2732 (1997)
23. T. Ditmire et al., Nature (London) **386**, 54 (1997)
24. Y.L. Shao et al., Phys. Rev. Lett. **77**, 3343 (1996)
25. A. McPherson et al., Nature **370**, 631 (1994)
26. T. Ditmire et al., Phys. Rev. Lett. **75**, 3122 (1995)
27. T. Ditmire et al., Phys. Rev. A **53**, 3379 (1996)
28. Y. Fukuda et al., Phys. Rev. Lett. **103**, 165002 (2009)
29. A.B. Borisov et al., J. Phys. B At. Mol. Opt. Phys. **36**, 3433 (2003)
30. T. Ditmire et al., Nature **398**, 489 (1999)
31. V. Kumarappan et al., Phys. Rev. Lett. **94**, 205004 (2005)

32. B. Shim et al., Phys. Rev. Lett. **98**, 123902 (2007)
33. S. Dobosz et al., Phys. Rev. A **56**, R2526 (1997)
34. M. Lezius et al., Phys. Rev. Lett. **80**, 261 (1998)
35. J. Zweiback et al., Phys. Rev. A **59**, R3166 (1999)
36. H. Wabnitz et al., Nature **420**, 482 (2002)
37. T. Laarmann et al., Phys. Rev. Lett. **92** (2004)
38. T. Laarmann et al., Phys. Rev. Lett. **95**, 063402 (2005)
39. C. Bostedt, et al., Phys. Rev. Lett. **100** (2008)
40. H. Thomas, et al., J. Phys. B At. Mol. Opt. Phys. **42**, 134018 (2009)
41. O.F. Hagena, W. Obert, J. Chem. Phys. **56**, 1793 (1972)
42. O.F. Hagena, Rev. Sci. Instrum. **63**, 2374 (1992)
43. K. Mendham et al., Phys. Rev. A **64**, 055201 (2001)
44. J. Wormer et al., Chem. Phys. Lett. **159**, 321 (1989)
45. F. Dorchies et al., Phys. Rev. A **68**, 23201 (2003)
46. I. Last, J. Jortner, Phys. Rev. A **60**, 2215 (1999)
47. L.V. Keldysh, Sov. Phys. JETP **20**, 1307 (1965)
48. W. Lotz, Z. Phys. **216**, 241 (1968)
49. V.P. Krainov, J. Phys. B At. Mol. Opt. Phys. **33**, 1585 (2000)
50. U. Saalmann, J.M. Rost, Phys. Rev. Lett. **91**, 223401 (2003)
51. U. Saalmann et al., J. Phys. B At. Mol. Opt. Phys. **39**, R39+ (2006)
52. N.W. Ashcroft, N.D. Mermin, *Solid State Physics* (Saunders College Publishing, Orlando, 1976)
53. H.M. Milchberg et al., Phys. Rev. E **64**, 056402 (2001)
54. A.R. Holkundkar, N.K. Gupta, Phys. Plasmas **15**, 013105 (2008)
55. J. Liu et al., Phys. Rev. A **64**, 033426 (2001)
56. F. Megi et al., J. Phys. B At. Mol. Opt. Phys. **36**, 273 (2003)
57. S. Micheau et al., High Energy Density Phys. **3**, 191 (2007)
58. P. Hilse et al., Laser Phys. **19**, 428 (2009)
59. J.C. Stewart, J.K.D. Pyatt, Astrophys. J. **144**, 1203 (1966)
60. P. Mulser et al., Phys. Rev. Lett. **95**, 103401 (2005)
61. P. Mulser, M. Kanapathipillai, Phys. Rev. A **71**, 063201 (2005)
62. I. Last, J. Jortner, Phys. Rev. A **62** (2000)
63. K. Ishikawa, T. Blenski, Phys. Rev. A **62**, 063204 (2000)
64. G. M. Petrov et al., Phys. Rev. E **71** (2005)
65. G.M. Petrov et al., Phys. Plasmas **12**, 063103 (2005)
66. G.M. Petrov, J. Davis, Eur. Phys. J. D **41**, 629 (2007)
67. M. Eloy et al., Phys. Plasmas **8** (2001)
68. Y. Kishimoto et al., Phys. Plasmas **9**, 589 (2002)
69. T. Taguchi et al., Phys. Rev. Lett. **92**, 205003 (2004)
70. C. Jungreuthmayer et al., Phys. Rev. Lett. **92**, 133401 (2004)
71. J. Barnes, P. Hut, Nature **324**, 446 (1986)
72. A.R. Holkundkar et al., Phys. Plasmas **18** (2011)
73. G. Mishra et al., Laser Part. Beams **29**, 305 (2011)
74. G.M. Petrov, J. Davis, Phys. Plasmas **15**, 056705 (2008)
75. I. Last, J. Jortner, Phys. Rev. A **75**, 042507 (2007)
76. M.V. Ammosov et al., Sov. Phys. JETP **64**, 1191 (1986)
77. G. Voronov, At. Data. Nucl. Data Table **65**, 1 (1997)
78. E. Springate et al., Phys. Rev. A **61**, 063201 (2000)
79. V. Kumarappan et al., Phys. Rev. Lett. **87**, 085005 (2001)
80. V. Kumarappan et al., Phys. Rev. A **66**, 33203 (2002)
81. M. Krishnamurthy et al., Phys. Rev. A **69**, 33202 (2004)
82. D. Mathur et al., Phys. Rev. A **82**, 025201 (2010)
83. D. Mathur, F.A. Rajgara, J. Chem. Phys. **133**, 061101 (2010)
84. E. Skopalová et al., Phys. Rev. Lett. **104**, 203401 (2010)

85. J. Kou et al., J. Chem. Phys. **112**, 5012 (2000)
86. M. Krishnamurthy et al., J. Phys. B At. Mol. Opt. Phys. **39**, 625 (2006)
87. G. Mishra, N.K. Gupta, Europhys. Lett. **96**, 63001 (2011)
88. G. Mishra, N.K. Gupta, Phys. Plasmas **19**, (2012)
89. T. Brabec, F. Krausz, Rev. Mod. Phys **72**, 545 (2000)
90. L. Xu et al., Opt. Lett. **21**, 2008 (1996)
91. D.J. Jones et al., Science **288**, 635 (2000)
92. A. Apolonski et al., Phys. Rev. Lett. **85**, 740 (2000)
93. A. Baltuska et al., Nature **421**, 611 (2003)
94. G.G. Paulus et al., Phys. Rev. Lett. **91**, 253004 (2003)
95. P.B. Corkum, Phys. Rev. Lett. **71**, 1994 (1993)
96. X. Liu, C.F. de Morisson, Faria. Phys. Rev. Lett. **92**, 133006 (2004)
97. G. Mishra, N.K. Gupta, J. Phys. B At. Mol. Opt. Phys. **46**, 125602 (2013)
98. D. Bauer, Laser Part. Beams **21**, 489 (2003)
99. J. Zweiback et al., Phys. Rev. Lett. **85**, 3640 (2000)
100. J. Zweiback et al., Phys. Rev. Lett. **84**, 2634 (2000)
101. K. Madison et al., Phys. Rev. Lett. **70**, 053201 (2004)
102. I. Last, J. Jortner, Phys. Rev. Lett. **87**, 033401 (2001)
103. J. Davis et al., Phys. Plasmas **13**, 064501 (2006)
104. A.R. Holkundkar et al., Phys. Plasmas **21**, 013101 (2014)
105. J.D. Huba, *NRL Plasma Formulary* (Naval Research Laboratory, Washington, D.C., 2009)
106. J.F. Ziegler et al., Nucl. Instrum. Meth. B **268**, 1818 (2010)

Chapter 5
Backward Lasing of Femtosecond Plasma Filaments

Yi Liu, Sergey Mitryukovskiy, Pengji Ding, Guillaume Point, Yohann Brelet, Aurélien Houard, Arnaud Couairon and André Mysyrowicz

Abstract Stimulated emissions in both backward and forward directions from a plasma filament in ambient air or pure nitrogen have been observed in recent years. In this article, we present our recent experimental results concerning the backward stimulated emission. We first demonstrate that backward stimulated emission from neutral N_2 molecules can be effectively generated with a circularly polarized 800 nm femtosecond laser pulse in pure nitrogen. Then, we show that the presence of oxygen is detrimental to the laser gain. To further confirm the presence of population inversion, we send a counter-propagating seeding pulse into the plasma filament. This leads to an amplification of the seeding pulse by two orders of magnitude. The crucial role of pump laser polarization indicates that the inelastic collisions between the energetic electrons and the neutral N_2 molecules are at the origin of population inversion between the relevant states.

Y. Liu · S. Mitryukovskiy · P. Ding · G. Point · Y. Brelet · A. Houard · A. Mysyrowicz (✉)
Laboratoire d'Optique Appliquée, ENSTA-Paristech/CNRS/Ecole Polytechnique,
828, Chemin des Maréchaux, 91762 Palaiseau, France
e-mail: andre.mysyrowicz@ensta-paristech.fr

Y. Liu
e-mail: yi.liu@ensta-paristech.fr

S. Mitryukovskiy
e-mail: SergeyMitryukovskiy@ensta-paristech.fr

P. Ding
e-mail: pengji.ding@ensta-paristech.fr

G. Point
e-mail: guillaume.point@ensta-paristech.fr

Y. Brelet
e-mail: yohann.brelet@ensta-paristech.fr

A. Houard
e-mail: aurelien.houard@ensta-paristech.fr

A. Couairon
Centre de Physique Théorique, CNRS, Ecole Polytechnique, 91128 Palaiseau, France
e-mail: arnaud.couairon@cpht.polytechnique.fr

K. Yamanouchi et al. (eds.), *Progress in Ultrafast Intense Laser Science XII*,
Springer Series in Chemical Physics 112, DOI 10.1007/978-3-319-23657-5_5

5.1 Introduction

Lasing in ambient air or its major components (N_2 and O_2) when pumped by an intense ultrashort laser pulse has attracted much attention in recent years [1–17]. The generation of coherent radiation with such cavity-less laser sources holds great potential for remote sensing applications. The advantage of the backward stimulated radiation for remote sensing lies in the fact that it can bring information about pollutants towards the ground observer with a well defined directionality. This has to be compared to the omni-directionality of the fluorescence or scattered optical signal from the same pollutants. In addition, if the information is carried to the ground observer via coherent Stokes- or anti-Stokes Raman scattering (CARS), the signal is proportional to N^2, as compared to the N dependence of the fluorescence or spontaneous scattering process [18]. Here N is the number of pollutant molecules.

Up to now, two different schemes for backward stimulated radiation in air have been demonstrated experimentally, based on population inversion either of O atoms or N_2 molecules. In 2011, Dogariu et al. showed that a backward coherent emission at 845 nm can be achieved in air pumped by picosecond UV pulses [1]. The stimulated emission was attributed to optical dissociation of O_2 molecules followed by multiphoton excitation of atomic Oxygen. A serious limitation of this scheme is the significant absorption of 226 nm light by atmosphere, preventing its use for remote pollutant sensing. Backward stimulated emission from neutral nitrogen molecules inside a laser plasma filament was also suggested in 2003, based on the observed exponential increase of the backward UV emission as a function of the filaments length [5]. In 2012, D. Kartashov and coworkers focused an infrared femtosecond laser pulse (3.9 or 1.03 μm) inside a high pressure mixture of argon and nitrogen gas. Backward stimulated emissions at 337 and 357 nm were observed with an optimal argon gas pressure of 5 bar and nitrogen pressure of 0.5 bar [2]. The emissions at 337 and 357 nm have been identified as the transition between the third and second excited triplet states of neutral nitrogen molecules, i.e. $C^3\Pi_u \rightarrow B^3\Pi_g$. This transition has been traditionally referred to as the second positive system of nitrogen, see Fig. 5.1. The population inversion mechanism between the $C^3\Pi_u$ and $B^3\Pi_g$ states has been attributed to the traditional Bennet mechanism, where collisions transfer the excitation energy of argon atoms to molecular nitrogen [2, 19]. Unfortunately, this method cannot be applied for remote generation of backward lasing emission because of its requirement of high pressure argon gas ($p > 3$ bar).

During the same period, in 2011 Yao et al. reported forward stimulated radiation from N_2 molecular ions pumped by short and intense mid-infrared pulses [10]. The lasing radiation stems from the $B^2\Sigma_u^+ \rightarrow X^2\Sigma_g^+$ transition between the excited and ground cation states of nitrogen molecules (Fig. 5.1). After that, several different methods have been reported for the generation of forward lasing [11–17]. Up to now, the mechanism responsible for between the $B^2\Sigma_g^+$ and $X^2\Sigma_g^+$ state is still mysterious, which is beyond the scope of the current paper. In this chapter, we will concentrate on the backward stimulated lasing action from laser plasma filaments.

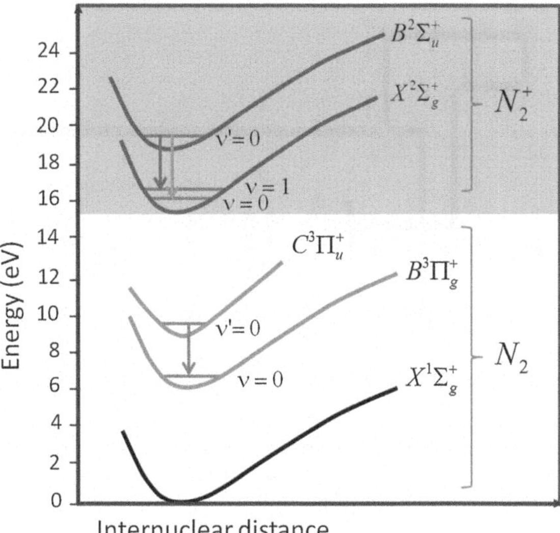

Fig. 5.1 Schematic potential energy diagram of the neutral and ionic nitrogen molecules. The $C^3\Pi_u^+ \rightarrow B^3\Pi_g^+$ transition of the neutral N_2 was historically named as the second positive band of nitrogen molecules, while the $B^2\Sigma_u^+ \rightarrow X^2\Sigma_g^+$ transition of the N_2^+ as the first negative band

We will first show the experimental evidences of a backward amplified spontaneous emission (ASE) obtained in pure nitrogen pumped with a circularly polarized 800 nm pulse. Experimental results concerning the radiation spectrum, spatial profile, polarization properties, and its dependence on incident laser pulse energy and ellipticity will be presented. We will then show that in the presence of an external seeding pulse, the backward stimulated emission can be enhanced by more than 20 times. In the mean time, the amplified emission inherits the polarization of the seeding pulse and exhibits a much smaller divergence than the backward ASE.

This article is organized as follows. In the first section, we introduce the background and the recent progress on lasing action of air. The experimental setup and the diagnostics will be briefly discussed in the second section. In the third section, the backward lasing emission from excited neutral N_2 inside plasma filaments will be discussed. Particular attention will be paid to the crucial role of pump laser polarization. We will present an external seeding scheme of backward lasing emission in the fourth section. A short summary will be given in Sect. 5.5.

5.2 Experimental Setup

In our experiment, a commercial Chirped-Pulse-Amplification laser system (Thales, Alpha 100) is used, which provides 45 fs pulses at 800 nm with energy up to 12 mJ per pulse. The output beam has a Gaussian profile with a diameter of 16 mm (1/e^2). A schematic experiment setup is presented in Fig. 5.2. The output laser pulse is split into a main pump pulse and a much weaker second pulse by a 1 mm thick 95 %/5 % beamsplitter. The pump pulse passes through a $\lambda/4$ waveplate and is then

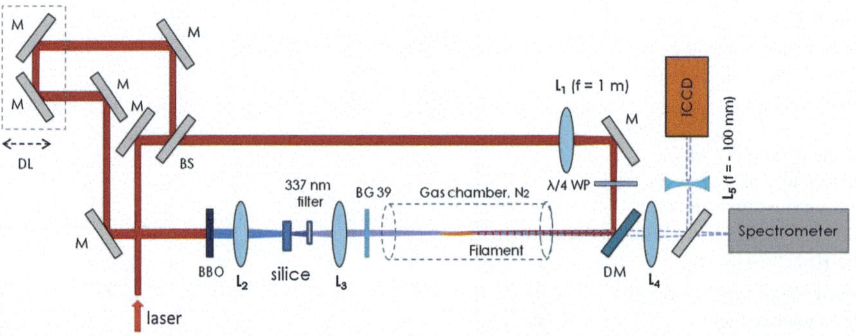

Fig. 5.2 Schematic experimental setup. The circularly polarized pump pulses (800 nm) is focused by a convex $f = 1000$ mm lens inside the gas chamber filled with pure nitrogen. The seed pulse at 337 nm is obtained from the supercontinuum generated inside a 2 cm long fused silica bulk by filamentation of the second harmonic (400 nm) pulses. The seed pulses are focused by an $f = 10$ cm lens in the opposite direction of the pump pulses, with its geometrical focus overlapping with the center of the pump filaments

focused by an $f = 1000$ mm convex lens (L1). A dichromatic mirror is used to reflect the focused 800 nm pump pulses into a gas chamber filled with pure nitrogen gas at 1 bar pressure. This dichromatic mirror reflects more than 99 % of the 800 nm pump pulse and it is transparent for the backward ultraviolet emission from the laser plasma situated inside the gas chamber. The second weaker 800 nm pulse first passes through a mechanical delay line and then a 1 mm thick type-I BBO crystal in order to generate femtosecond pulses at 400 nm. The 400 nm pulses are linearly polarized in the vertical direction. The obtained 400 nm pulses are further focused by an $f = 200$ mm convex lens (L2) inside a 20 mm long fused silica sample to broaden its spectrum through intense nonlinear propagation. We select the spectrum component around 337 nm with an interference bandpass filter, which has a transmission peak at 340 nm and a bandwidth of 10 nm. The resulting pulse centered at 340 nm, referred to as seeding pulse in the following, is focused by another $f = 100$ mm lens (L3) into the gas chamber from the opposite direction of the pump pulses. The separation between the lenses L2 and L3 is adjusted to insure that the geometrical focus of the seeding pulse overlaps with the central part of the long plasma filament in the longitudinal direction. The transverse spatial overlapping between the geometrical focus of the seeding pulse and the center of the plasma filament is carefully ensured by translating finely the focusing lens (L3). The temporal delay between the 800 nm pump pulses and the seeding pulse at 337 nm can be adjusted by the mechanical delay line. For some experiments, a $\lambda/4$ waveplate for 400 nm is installed after the BBO crystal, where circularly polarized seeding pulse can be obtained after filamentation inside the fused silica sample. The backward emission from the laser plasma filaments is detected by either a spectrometer (Ocean Optics HR 4000), an intensified Charge Coupled Camera (Princeton Instrument, PI-MAX), or a calibrated photodiode.

5.3 Backward Lasing Emission from Laser Plasma Filaments

In this section, we first discuss the backward ASE from laser filament in pure nitrogen and its mixture with oxygen. It is highlighted that the backward ASE can be only obtained with circularly polarized laser pulses at 800 nm. The possible mechanism for population inversion will be addressed at the end of this subsection.

5.3.1 Backward Stimulated UV Emission from Filaments in Nitrogen Gas

We first examine the spontaneous UV fluorescence emitted by filaments in a pure Nitrogen gas. The fluorescence spectra recorded in the transverse direction are shown in Fig. 5.3a for linearly and circularly polarized pump laser pulses. The emission peaks centered at 315, 337, 357, 380, 405 nm have been previously identified as due to transitions between various vibronic levels of the triplet $C^3\Pi_u$ and $B^3\Pi_g$ states of the neutral N_2 molecule, i.e. the second positive band of the N_2 molecules [19]. For all these five spectral lines, the signals are $\sim$2 times stronger with circularly polarized laser than with linear laser polarization. A possible explanation will be discussed at the end of this subsection.

We now concentrate on the backward emission at 337 nm obtained with circularly polarized femtosecond laser pulses. It corresponds to the (0–0) vibronic transition of the second positive band system of N_2 molecule. In Fig. 5.3b, the spectra of the backward UV emission are shown for circular and linear polarization of the laser. The emission intensity at 337 nm is now about 10 times larger with circular pump laser polarization than with linear polarization. The remarkable behavior of the 337 nm

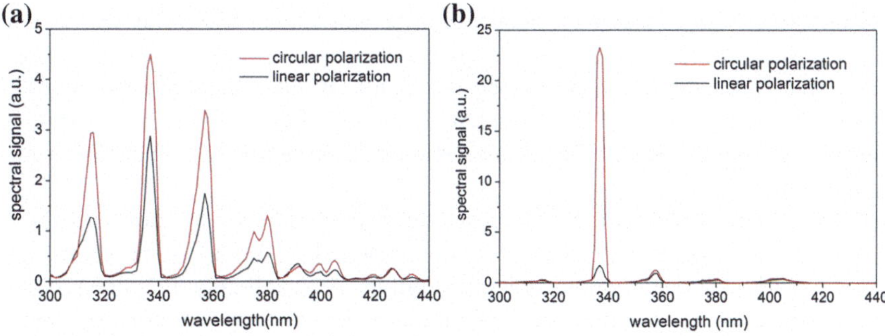

Fig. 5.3 Spectra of **a** transverse fluorescence and **b** backward UV emission for circular and linear laser polarization. The incident laser pulse energy is 9.3 mJ

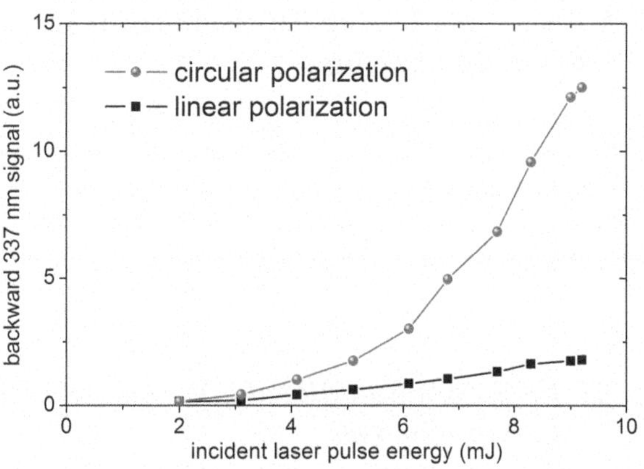

Fig. 5.4 Measured backward signal at 337 nm as a function of incident laser pulse energy, for both circular and linear laser polarization

signal suggests that backward stimulated emission is initiated for this particular line with circularly polarized laser pulses.

In Fig. 5.4, the backward emission intensity at 337 nm is plotted as a function of incident laser energy, for both circularly and linearly polarized pump pulses. The 337 nm signal for circular laser polarization displays a superlinear dependence on incident laser energy, which is much higher than that of linearly polarized pump lasers for pump laser energy higher than 5 mJ.

To get further insight into the distinct pump polarization dependence presented in Fig. 5.4, we measured the backward radiation intensity at 337 nm as a function of the incident laser ellipticity. In Fig. 5.5, the measured signals are presented as a function of the rotation angle of the quarter-wave plate for different incident laser energies. The angles $\varphi = 90° \times m$ correspond to linearly polarized laser, with $m = 0, 1, 2, 3$. The angles $\varphi = 45° + 90° \times m$ correspond to circularly polarized laser. For a low pulse energy of 300 μJ, linearly polarized pulses generate a UV radiation with an intensity twice that obtained with circular polarization (Fig. 5.5a). For an increased incident energy of 600 μJ, an octagon-shaped dependence is observed (Fig. 5.5b), indicating the onset of a new mechanism for the emission at 337 nm. In the case of $E_{in} = 9.3$ mJ, the signal obtained with circular laser polarization totally dominates and decreases rapidly when the laser polarization deviates slightly from circular. This critical dependence of the emission signal at 337 nm with laser polarization reinforces the hypothesis that backward stimulated emission at 337 nm occurs inside the filament plasma.

Finally, we present the measurements of the spatial profile of the backward emission. In Fig. 5.6, the profiles of the emission captured by the ICCD are shown for both circularly and linearly polarized pump pulses. A clear highly directional beam profile, with a divergence angle of ~9.2 mrad, can be observed for circularly polarized pump pulses. For linearly polarized pump pulses, no observable beam can be detected by the ICCD. These spatial profile measurements agree with the above spectral measurements where significant stimulated emission at 337 nm can only be observed with circularly polarized pump pulses.

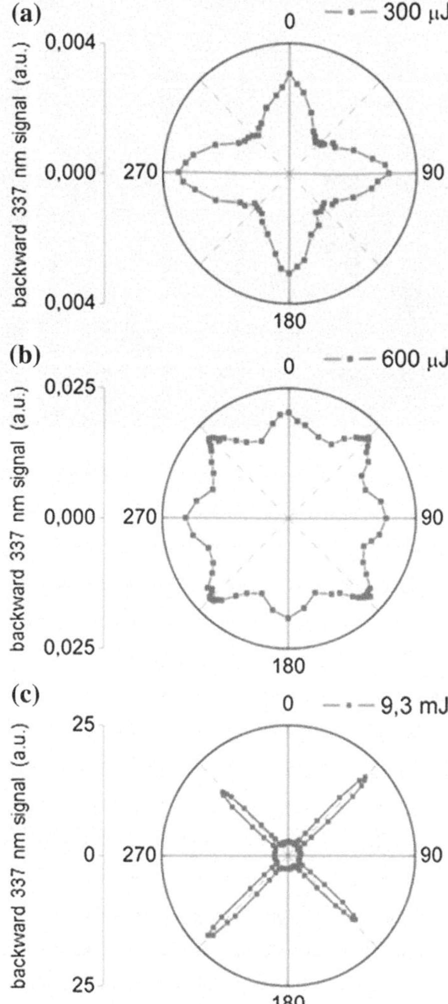

Fig. 5.5 Backward emission signal at 337 nm as a function of the rotation angle of the quarter-wave plate. The incident laser energy for **a**, **b**, **c** are 0.3, 0.6, and 9.3 mJ, respectively. Angle 0° corresponds to linearly polarized light

5.3.2 Forward ASE from Filaments in Pure Nitrogen

We also tried to observe the stimulated emission at 337 nm in the forward direction. In pure nitrogen, intense radiation peaked around 337 nm was observed for circularly polarized 800 nm pulses. The measured forward emission spectra are presented in Fig. 5.7a for both circularly and linearly polarized pump pulses. In the case of linear laser polarization, no observable forward ASE at 337 nm was detected. In Fig. 5.7b, we presented the intensity of the forward 337 nm radiation as a function of the rotation angle of the $\lambda/4$ waveplate, where a similar dependence to that of backward ASE (Fig. 5.5c) was observed.

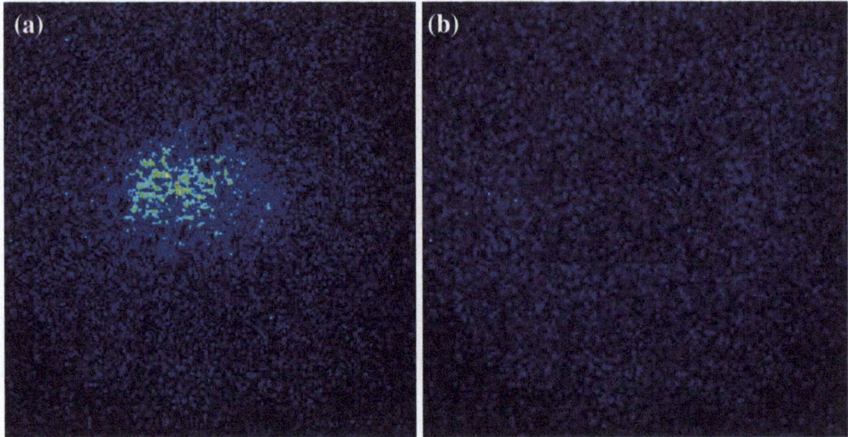

Fig. 5.6 Profile of the backward emission at 337 nm with circularly (**a**) and linearly (**b**) polarized femtosecond pump pulses at 800 nm

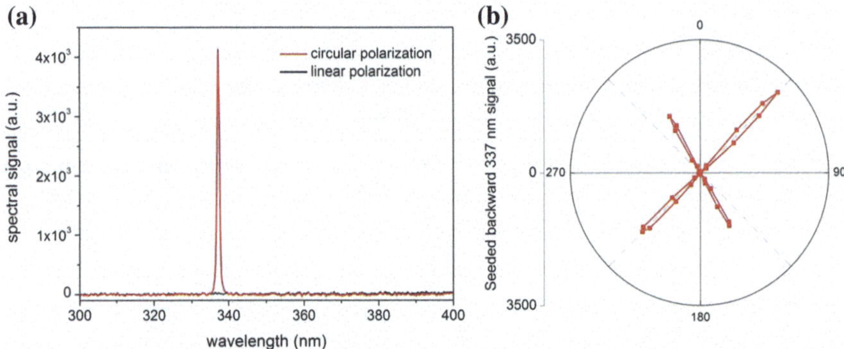

Fig. 5.7 **a** Forward emission spectrum from the filaments in pure nitrogen pumped by circularly and linearly polarized laser pulses. The focal length was $f = 1000$ mm and the incident pulse energy was 9.3 mJ. **b** Forward 337 nm intensity as a function of the rotation angle of the $\lambda/4$ waveplate for the 800 nm pump pulses

5.3.3 Backward Emission at 337 nm from Filament in Air

We noticed that with the addition of Oxygen gas to the nitrogen gas, the backward 337 nm emission signal decreases gradually. We present in Fig. 5.8 the measured 337 nm signal for both circularly and linearly polarized laser as a function of the percentage of Oxygen. A slow decrease of the signal is observed upon increasing Oxygen concentration up to 10 %. Beyond 10 % Oxygen, the signal shows a rapid decrease, indicating the termination of significant lasing action. We also confirmed with the ICCD that no detectable ASE radiation was observed in ambient air in the case of circularly polarized pulses.

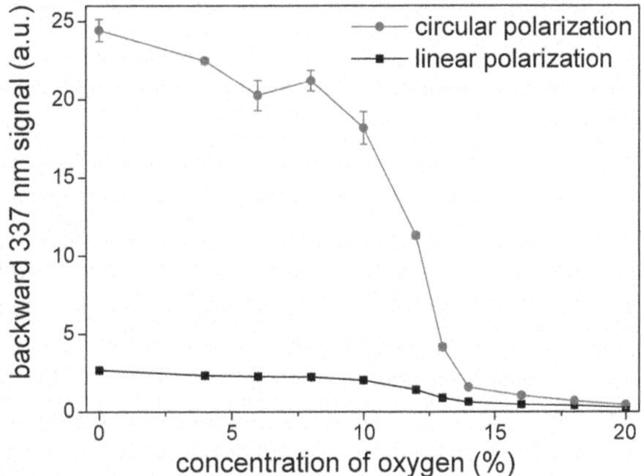

Fig. 5.8 Measured backward 337 nm signal from filaments in air as a function of oxygen gas concentration, for both circular and linear polarized laser pulses. The incident pulse energy is 9.3 mJ

The detrimental influence of Oxygen molecule for the conventional discharge-pumped Nitrogen laser is well documented [19]. The underlying physical mechanism is attributed to the collision reaction

$$N_2(C^3\Pi_u^+) + O_2 = N_2(X^1\Sigma_g^+) + O + O,$$

which efficiently reduces the population density in the upper state of the lasing emission.

To obtain a significant backward stimulated emission in atmospheric air, a higher population inversion density (or a higher gain) is required to overcome the quenching effect of the Oxygen molecules. A possible approach to achieve this is to use a pump laser at longer wavelengths, because the kinetic energy of the produced electrons increases like $I\lambda^2$, where I is the laser intensity and λ the wavelength.

5.3.4 Discussion of the Population Inversion Mechanism

We now discuss possible mechanisms for population inversion between the $C^3\Pi_u$ and $B^3\Pi_g$ states of the neutral N_2 molecule, at the origin of the backward stimulated radiation. First, it is worth reminding that a direct population transition between the ground singlet state $X^1\Sigma_g^+$ and the excited triplet $C^3\Pi_u^+$ state is forbidden in the electric dipole approximation. A widely discussed mechanism to populate $C^3\Pi_u^+$ state inside filaments in air is the following reaction [20]:

$$N_2^+ + N_2 = N_4^+,$$
$$N_4^+ + e = N_2^+(C^3\Pi_u^+) + N_2(X^1\Sigma_g^+).$$

With this mechanism, one expects that a linearly polarized laser produces a stronger fluorescence signal for two reasons. First, it is known that the ionization rate of atoms and molecules is higher for linear polarization both in the multiple photon ionization and tunneling ionization regime [21]. As a result, the densities of electrons and positive ions N_2^+ are higher in the case of linear laser polarization and should result in a higher density of $N_2^+(C^3\Pi_u^+)$. Second, the nonlinear refractive index of ambient air n_2 is higher for linear polarized laser [21]. This results in slightly higher laser intensity inside filaments, which should again lead to higher densities of electrons and ions. We therefore assume that this collision-assisted recombination process of the electron on the parent ion is responsible for the fluorescence of filaments at relatively low laser energy, such as that of $E_{\text{in}} = 300\ \mu\text{J}$ in Fig. 5.6a, but not for the stimulated emission observed at higher pump powers.

Another mechanism for the transition from the ground state to the $C^3\Pi_u^+$ state is the electron-molecule inelastic collision:

$$N_2(X^1\Sigma_g^+) + e = N_2(C^3\Pi_u^+) + e.$$

This is actually the main reaction responsible for population inversion in a traditional N_2 laser, where the electrons are accelerated to obtain sufficient energy by the discharge electric field [19]. The cross section of the above reaction is sensitive to the kinetic energy of the incident electron. It is nearly zero for electron energy below the threshold kinetic energy $E_{th} \sim 10$ eV, exhibits a resonant peak around 14.1 eV, and then decreases progressively for higher energy electrons [22]. For electrons born in the intense laser field inside filaments, the distribution of kinetic energies depends strongly upon pump laser polarization. With linear polarization, the photoelectrons have a low kinetic energy, with a weak tail distribution extending to a few eV. For circularly polarized light, where electrons are always accelerated away from the ion core, the final electron distribution is nearly mono-kinetic with a peak at twice the ponderomotive energy. Calculations performed for a filament show that the peak of this distribution is between 5 and 7 eV assuming a laser intensity of 5×10^{13} W/cm^2 and should scale up linearly with laser intensity [23]. The widely quoted clamped intensity value in air filaments of 5×10^{13} W/cm^2 is therefore insufficient by a factor 2 to be an effective excitation mechanism. On the other hand, it is now well established that intensity clamping is not a rigorous concept. Intensity spikes exceeding the clamped value by order of magnitude have been measured in filaments [24, 25].

In our current experiment, we have estimated experimentally the laser intensity inside the filament to be around 1.45×10^{14} W/cm^2 [26, 27]. This indicates that most of the electrons can obtain kinetic energy of ~16 eV. In another words, the electron energy required for impact excitation of neutral N_2 molecules is well reached in filaments. For linearly polarized laser field, this mechanism is turned off because of the low electron kinetic energy, even with laser intensity above the clamped value.

5.4 External Seeded Backward Lasing from Filaments

To further confirm the existence of population inversion between the $C^3\Pi_u$ and $B^3\Pi_g$ states, we send a seeding pulse around 337 nm in the opposite direction with respect to the pump pulses (Fig. 5.2). The idea is that if there is population inversion, the seeding pulse should be significantly amplified and the amplified radiation should bear the physical properties of the seeding pulse, such as polarization.

We first measure the spectra of the backward emission from the laser plasma in the presence of the seeding pulse, for both circularly and linearly polarized pump lasers. The results are presented in Fig. 5.9. In the case of circularly polarized pump pulses, a sharp radiation peak at 337 nm can be observed without external seed pulse, which is the backward ASE discussed in the above subsection. With the external seed pulse, an enhancement of the peak by a factor of ~20 is found (Fig. 5.9a). Considering the intensity of the seed pulse at the 337 nm spectral position, we estimate that the seed pulse is amplified by a factor of $2.5 \times 10^5/2000 = 125$ times. In the case of linearly polarized pump pulses (Fig. 5.9b), no detectable ASE is observed and no amplification of the seed pulse can be observed. These observations confirm that population inversion responsible for the 337 nm stimulated radiation is only established with circularly polarized pump pulses.

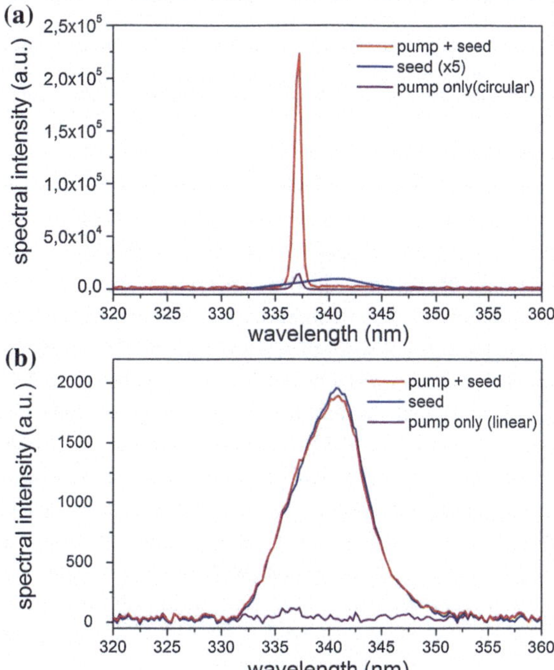

Fig. 5.9 Spectrum of the backward stimulated emission with circularly (**a**) and linearly (**b**) polarized pump pulsed at 800 nm. The spectra of the seed pulses and those of the backward emission from the filaments without seeding pulse are also presented for comparison

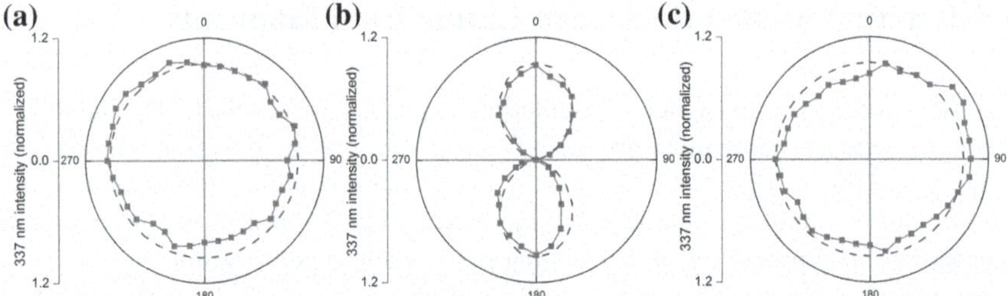

Fig. 5.10 Polarization properties of the ASE (**a**), the seeded stimulated backward emission with linearly (**b**) and circularly (**c**) polarized seed pulses at 337 nm. The *dots* present the experimental results and the *dashed lines* denote theoretical fitting

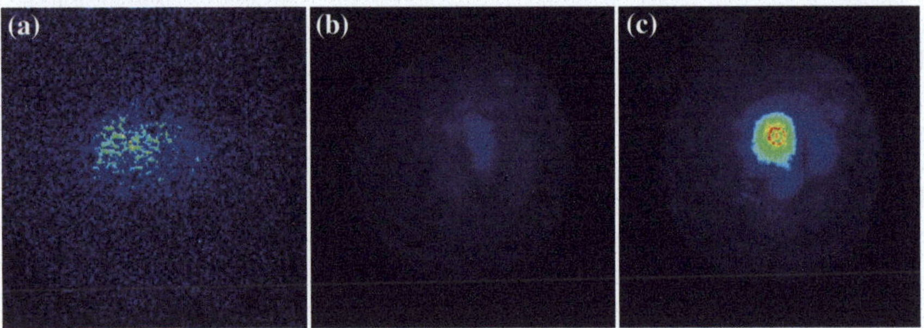

Fig. 5.11 Spatial profile of the backward ASE emission (**a**), the seed pulse (**b**), and the seeded stimulated radiation (**c**). The opening angle of each panel is 24 mrad × 24 mrad. The intensity in (**a**) is not in the same scale as those of (**b**) and (**c**) due to different filters being used

We then study the polarization property of this seeding effect by injecting linearly and circularly polarized seeding pulses inside the plasma filament. In order to analysis the polarization property of the lasing radiation, we install a Glan prism before the detecting photodiode. In the experiment, we record the intensity of the transmitted 337 nm radiation as a function of the rotation angle of the Glan prism. The result for the ASE obtained without seeding pulse is first presented in Fig. 5.10a. It is clear that the ASE is not polarized. For linearly polarized seed pulses in the vertical direction, we observe that the amplified lasing signal is also linearly polarized in the same direction (see Fig. 5.10b), evidenced by the good agreement between the experimental results and the theoretical fit with Malus' law. The result for circularly polarized seeding pulses is presented in Fig. 5.10c, where a circularly polarized amplified emission is also observed. The conservation of the pulse polarization during the amplification is in agreement with our hypothesis that that population inversion is present and responsible for the seeding pulse amplification.

We also measure the spatial profile of the ASE and the seeded lasing emission. In Fig. 5.11a, the profile of the backward ASE obtained with circularly polarized pump pulses is re-shown for easier comparison. In Fig. 5.11b, the spatial profile of the seeding pulse is presented. In the presence of both pump and seeding pulse, an

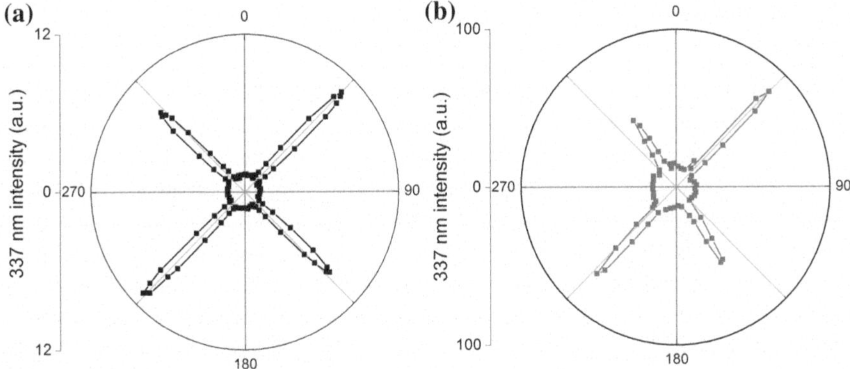

Fig. 5.12 Dependence of the backward ASE (**a**) and the seeded backward stimulated radiation (**b**) on the rotation angle of the quarter-wave plate

extremely intense 337 nm radiation is found, as presented in Fig. 5.11c. This amplified stimulated emission shows a divergence angle of ~4 mrad, much smaller than that of the seed pulse and the ASE.

All the above experimental observations confirm again the crucial role of pump laser polarization, which has been observed for the backward ASE in the above section. To evaluate the role of pump laser polarization, we measure systematically the seeded lasing emission intensity by rotating the $\lambda/4$ waveplate for the pump pulses. In Fig. 5.12a, the result for the ASE without seeding pulse is first presented as a function of the rotation angle ϕ of the waveplate, which has been reported in our previous work [8]. Intense ASE was observed only with circularly polarized pump pulses and shows dramatic decrease when the ellipicity deviates from $\varepsilon = 1$. In the presence of a constant linearly polarized seeding pulse, a similar dependence on laser ellipticity was observed (Fig. 5.12b). This confirms that population inversion between the $C^3\Pi_u$ and $B^3\Pi_g$ states can be only achieved with circularly polarized pump pulses. The slight asymmetry and the deviation of the peaks from $\varphi = 135°$ and $\varphi = 315°$ can be due to the fact that the circular polarized pump pulses reflect on the dielectric dichromatic mirror in this experiment. This mirror has slight different reflectivity for p- and s-polarized light and thus results in a non-perfect circularly polarized pump pulses after reflection.

Finally, we measure the pulse energy of the seeded lasing radiation with a sensitive laser energy meter. The maximum lasing energy was estimated to be 5 nJ. This corresponds to a conversion efficiency of $\sim 5 \times 10^{-7}$ from the 9.8 mJ pump pulse to the backward lasing emission.

In this seeded lasing scheme, we observed that no significant backward lasing signal can be achieved in ambient air, similar to the situation of backward ASE. This indicates that the detrimental role of Oxygen molecules is due to its ability to ruin population inversion, which was attributed to the collision quenching process [8].

5.5 Conclusion

In conclusion, we have shown that a strong stimulated radiation at 337 nm can be achieved in the backward direction from filament plasma in nitrogen gas pumped by a circularly polarized laser pulse at 800 nm. This stimulated radiation shows distinct dependence on laser pulse energy, compared to that obtained with linear polarized laser pulses. In a mixture of nitrogen gas and oxygen, the presence of Oxygen molecules suppresses the lasing action to a large extent. As to the mechanism responsible for population inversion, we attribute it to inelastic collisions between electrons liberated by the pump laser and neutral nitrogen molecules, a process which is more efficient with circularly polarized laser pulses. We believe that this simple scheme for backward stimulated emission from nitrogen gas pumped by the widely available 800 nm femtosecond laser pulse is a significant step towards applications for remote sensing.

To further confirm the existence of population inversion, we have injected a seeding pulse in the opposite direction of the pump pulse, which leads to an enhancement of the backward radiation intensity by a factor of ~20. The amplified lasing radiation inherits the polarization property of the seeding pulse and its divergence angle was found to be around 3.8 mrad, 3 times smaller than that of the backward ASE. This external seeding scheme provides a possible method to boost the energy of the backward lasing radiation, which is important for future applications.

Acknowledgments We gratefully acknowledge useful discussions with Paul Corkum of Ottawa University. The authors are also grateful to Thierry Lefrou and Aurélie Jullien of LOA for important technical help. Yi Liu acknowledges the stimulating discussion with Ya Cheng and Jinping Yao of SIOM, Huailiang Xu of Jilin University, Hongbing Jiang of Peking University, Benjamin Forestier of CILAS.

References

1. A. Dogariu, J.B. Michael, M.O. Scully, R.B. Miles, Science **331**, 442 (2011)
2. D. Kartashov, S. Ališauskas, G. Andriukaitis, A. Pugžlys, M. Shneider, A. Zheltikov, S.L. Chin, A. Baltuška, Phys. Rev. A **86**, 033831 (2012)
3. P.R. Hemmer, R.B. Miles, P. Polynkin, T. Siebert, A.V. Sokolov, P. Sprangle, M.O. Scully, Proc. Natl. Acad. Sci. USA **108**, 3130 (2011)
4. P. Sprangle, J. Peñano, B. Hafizi, D. Gordon, M. Scully, Appl. Phys. Lett. **98**, 211102 (2011)
5. P. Sprangle, J. Peñano, B. Hafizi, D. Gordon, R. Fernsler, J. Appl. Phys. **111**, 033105 (2012)
6. Q. Luo, W. Liu, S.L. Chin, Appl. Phys. B **76**, 337 (2003)
7. S. Owada, A. Azarm, S. Hosseini, A. Iwasaki, S.L. Chin, K. Yamanouchi, Chem. Phys. Lett. **581**, 21 (2013)
8. S. Mitryukovskiy, P. Ding, A. Houard, A. Mysyrowicz, Y. Liu, Opt. Express, **22**, 12750 (2014)
9. P. Ding, S. Mitryukovskiy, A. Houard, A. Couairon, A. Mysyrowicz, Y. Liu, Opt. Express, **22**, 29964 (2014)
10. J. Yao, B. Zeng, H. Xu, G. Li, W. Chu, J. Ni, H. Zhang, S.L. Chin, Y. Cheng, Z. Xu, Phys. Rev. A **84**, 051802(R) (2011)
11. J. Yao, G. Li, C. Jing, B. Zeng, W. Chu, J. Ni, H. Zhang, H. Xie, C. Zhang, H. Li, H. Xu, S.L. Chin, Y. Cheng, Z. Xu, New J. Phys. **15**, 023046 (2013)

12. Y. Liu, Y. Brelet, G. Piont, A. Houard, A. Mysyrowicz, Opt. Express **21**, 22791 (2013)
13. T. Wang, J. Ju, J.F. Daigle, S. Yuan, R. Li, S.L. Chin, Las. Phys. Lett. **10**, 125401 (2013)
14. H. Zhang, C. Jing, J. Yao, G. Li, B. Zeng, W. Chu, J. Ni, H. Xie, H. Xu, S.L. Chin, K. Yamanouchi, Y. Cheng, Z. Xu, Phys. Rev. X **3**, 041009 (2013)
15. W. Chu, G. Li, H. Xie, J. Ni, J. Yao, B. Zeng, H. Zhang, C. Jing, H. Xu, Y. Cheng, Z. Xu, Las. Phys. Lett. **10**, 125401 (2013)
16. G. Point, Y. Liu, Y. Brelet, S. Mitryukovskiy, P.J. Ding, A. Houard, A. Mysyrowicz, Opt. Lett. **39**, 1725 (2014)
17. D. Kartashov, S. Ališauskas, A. Baltuška, A. Schmitt-Sody, W. Roach, P. Polynkin, Phys. Rev. A **88**, 041805 (R) (2013)
18. P.N. Malevich, D. Kartashov, Z. Pu, S. Alisauskas, A. Pugzlys, A. Baltuska, L. Ginniunas, R. Danielius, A.A. Lanin, A.M. Zheltikov, M. Marangoni, G. Cerullo, Opt. Express **20**, 18784 (2012)
19. R.S. Kunabenchi, M.R. Gorbal, M.I. Savadatt, Prog. Quantum Electron. **9**, 259 (1984)
20. H.L. Xu, A. Azarm, J. Bernhardt, Y. Kamali, S.L. Chin, Chem. Phys. **360**, 171 (2009)
21. M. Kolesik, J.V. Moloney, E.M. Wright, Phys. Rev. E **64**, 046607 (2001)
22. J.T. Fons, R.S. Schappe, C.C. Lin, Phys. Rev. A **53**, 2239 (1996)
23. B. Zhou, A. Houard, Y. Liu, B. Prade, A. Mysyrowicz, A. Couairon, P. Mora, C. Smeenk, L. Arissian, P. Corkum, Phys. Rev. Lett. **106**, 255002 (2011)
24. E. Schulz, D.S. Steingrube, T. Binhammer, M.B. Gaarde, A. Couairon, U. Morgner, M. Kovačev, Opt. Express **19**, 19495 (2011)
25. M.B. Gaarde, A. Couairon, Phys. Rev. Lett. **103**, 043901 (2009)
26. Y. Liu, M. Durand, A. Houard, B. Forestier, A. Couairon, A. Mysyrowicz, Opt. Commun. **284**, 4706 (2011)
27. S. Mitryukovskiy, Y. Liu, A. Houard, A. Mysyrowicz, J. Phys. B: At. Mol. Opt. Phys. **48**, 094003 (2015)

Chapter 6
Propagation of Ultrashort, Long Wavelength Laser Pulses

Jayashree A. Dharmadhikari, Deepak Mathur and Aditya K. Dharmadhikari

Abstract This chapter summarizes some recent work on the propagation of long wavelength, intense, ultrashort laser pulses through different media. The spectral extent and spectral profile of the supercontinuum (SC) generated in transparent solids like BaF_2, CaF_2 and fused silica in the normal and anomalous group velocity dispersion regime (AGVD) are explored. Propagation of ultrashort pulses results in formation of filaments and gives rise to SC generation which is distinctly different in the AGVD regime compared to the normal GVD regime. A signature of SC generation in the AGVD regime is the appearance of an isolated wing on the blue-side of the spectrum in fused silica and CaF_2. Fifteen-photon absorption induced fluorescence is utilized to visualize femtosecond filaments in BaF_2 crystals and its absorption cross-section is deduced to be 6.5×10^{-190} cm^{30} W^{-15} s^{-1}. Water shows significant linear absorption in the anomalous dispersion regime. Results of a systematic study on SC generation in water, carried out in different GVD regimes at longer wavelengths, highlights the robustness of SC generation despite the presence of linear absorption in water. Long wavelength propagation facilitates generation of lower order harmonics (fifth and seventh) and permits the testing of the validity of higher-order Kerr effects in air. A recent advance in application of long wavelength filamentation in biological media is described, specifically relating to DNA damage induced at long wavelengths.

6.1 Introduction

Propagation of ultrashort laser pulses through transparent media is accompanied by generation of spectrally broad radiation called the supercontinuum (SC). SC generation has continued to be a subject of contemporary interest [1, 2] ever since its

J.A. Dharmadhikari
Department of Atomic and Molecular Physics, Manipal University, Manipal 576 104, India
e-mail: jayadh@gmail.com

D. Mathur · A.K. Dharmadhikari (✉)
Tata Institute of Fundamental Research, 1 Homi Bhabha Road, Mumbai 400 005, India
e-mail: atmol1@tifr.res.in; aditya@tifr.res.in

K. Yamanouchi et al. (eds.), *Progress in Ultrafast Intense Laser Science XII*,
Springer Series in Chemical Physics 112, DOI 10.1007/978-3-319-23657-5_6

observation in 1970 by Alfano and co-workers who focused picosecond laser pulses inside condensed media [3]. In 1986 spectral super-broadening was observed by Corkum et al. in gases using 70 fs UV pulses and 2 ps dye laser pulses [4]. Braun et al. observed very long filaments in air with fs laser without making use of a focusing lens [5]. It was soon appreciated that the high peak powers associated with ultrashort pulses inevitably give rise to a host of nonlinear optical effects in such media. Since then efforts have been on to study SC generation and filamentation in diverse media.

SC generation is a source of coherent broadband light that finds applications in various fields, such as pump-probe spectroscopy, white light microscopy, carrier-envelope phase stabilization, optical frequency comb [6], and as a seed for optical parametric amplification of ultrashort pulses. Considerable work has been reported in recent years on SC generation in solid media as a function of laser energy, polarization, pulse duration, and focusing conditions, but almost all experimental work has been carried out at 800 nm central wavelength, and under conditions of normal group velocity dispersion (GVD) [1].

Filamentation is a visual manifestation of the propagation of ultrashort, high-intensity pulses through matter. In filamentation manifestation of the optical Kerr effect results in self-focusing that, in turn, enables beams to propagate over distances much larger than the normal Rayleigh range. The self-focusing cannot continue indefinitely and is arrested by various mechanisms. In condensed media these mechanisms include diffraction, group velocity dispersion, self-phase modulation, and pulse self-steepening, higher-order nonlinear effects ($\chi^{(5)}$effects), and defocusing induced by plasma formation; all these factor contribute to the complexity of the overall propagation dynamics. Multiphoton absorption is an alternate mechanism for the arrest of self-focusing [7].

Propagation over long distances, which is a consequence of the dynamic competition between the optical Kerr effect and multiphoton ionization, results in a chain of focusing-refocusing events. A model has been proposed by Dubietis et al. [8] that do not require plasma generation, without ruling out the possibility that electrons are generated. A self-guiding effect in condensed media has been theoretically investigated by Skupin et al. [9]. At very high laser powers multiple filamentation occurs which scales with laser power [10]. The factors that affect the filament length and plasma density are the initial pulse intensity, initial pulse duration, and beam convergence [11]. Both plasma and GVD induced pulse splitting lead to pulse self-shortening, with a single compressed pulse component emerging at some distance, followed by rapid deterioration [12]. As a result, it becomes difficult to achieve a controlled and stable filamentation regime with laser pulse features preserved over extended distances.

In earlier studies carried out in our laboratory with 800 nm pulses (NGVD regime) in barium fluoride (BaF_2) we demonstrated highly efficient white light generation and systematic measurements of the spectral extent of the supercontinuum under different experimental conditions, such as laser energy, polarization, pulse duration, and lens focal length [13]. At very high incident powers ($\sim 3000P_{cr}$) we estimated the time-varying change of electron densities [14]. Some measure of control on the onset of filamentation within a large BaF_2 crystal was demonstrated [15] and the six-photon absorption cross-section estimate was obtained [16].

In recent years due to ready availability of femtosecond sources at longer wavelengths, theoretical and experimental efforts have recently begun to investigate filamentation and SC generation in AGVD regime [12, 17–22]. In the anomalous group velocity dispersion regime, one would expect that the self-guided pulses do not spread in time and space over long distances [23]. Anomalous group velocity dispersion will also induce a transfer of energy into the collapse region and will result in a substantial increase of the characteristic distances at which multiple collapses occur; the collapse region is substantially extended for anomalous GVD [18]. Measurements carried out in a 30 cm long BK7 sample showed that the length of the filament in the anomalous-GVD regime was as much as ~10 times longer than that in the normal-GVD regime [22]. Moreover in the multiple-collapse events, the separation distance was signficantly longer in the anomalous-GVD regime. In the normal-GVD regime, collapse occurs along the transverse spatial dimensions, whereas, in the anomalous-GVD regime, collapse occurs in both the spatial and temporal dimensions. Plasma formation- which is the main defocusing mechanism—arrests the collapse by lowering the refractive index and absorbing the pulse. In the case of normal GVD, energy is transferred rapidly away from the collapsing region, and this causes defocusing. In the anomalous-GVD regime, after plasma formation arrests collapse, the anomalous GVD may continue to transfer energy into the collapse region, resulting in the formation of longer filaments before the beam eventually defocuses [22].

In the anomalous group velocity dispersion region, red frequencies have a slower velocity than blue components. The generated frequencies are swept back into the main pulse. As a consequence, one expects a lower threshold for filamentation and more stable filamentary pulses in the region of negative group velocity dispersion (GVD). Thus, the filamentation process in the anomalous region leads to the formation of a pulse that is almost invariant with propagation and propagates as a spatiotemporal soliton. Of course, the pulse is lossy as it consumes energy for ionization of the medium; part of the light energy is dissipated in the form of a conical emission [24].

Numerical simulations have been reported on the propagation of longer wavelength laser pulses in fused silica (SiO_2) [22, 25, 26]. Measurements in fused silica show that the threshold for filamentation scales with wavelength in the 1200–2400 nm range, which is consistent with the fact that critical power for self-focusing also increases with the square of the wavelength [18]. A broadband emission is formed which can lead to efficient pulse self-compression [17, 27, 28]. In sapphire a tunable source (1100–1600 nm) has been used to produce a broader spectrum compared to that obtained with 800 nm pumping [29]. Recent reports on SC generation at longer wavelengths reconfirm the observation of a broader continuum, spanning more than 3 octaves, that could self-compress to a single-cycle pulse in the anomalous dispersion regime [18, 30–33]. Pulse self-compression in both the normal dispersion [34] and the anomalous dispersion regimes with 3.1 μm wavelength have been reported [35]. Porras et al. [36, 37] have interpreted measurements of filamentation and conical emission in the normal and anomalous dispersion regimes as a manifestation of the nonsolitary stationary solutions to the nonlinear Schrödinger equation. These solutions, called nonlinear X- or O-waves, arise from the self-focusing dynamics in Kerr media with nonlinear losses in normal or anomalous dispersion, respectively.

Experiments carried out in fused silica at 1.5 μm wavelength (in the anomalous GVD regime) showed a broad maximum around 600 nm, with the spectral extent covering the range 400–950 nm [12, 17]. An extreme blue-shifted continuum peak in the visible region was observed even though the filament was formed using incident pulses that were in the near-IR [38]. The blue-side peak was identified as an axial component of the conical emission, as proposed in a three-wave mixing model [18]. Self-phase modulation undergoes strong self-steepening in AGVD regime. This leads to greater intensity gradient at the tailing edge of the pulse causing larger broadening. The blue side peak was also identified as an isolated anti-Stokes wing (ASW) which is formed by the interference of the SC light field encountering anomalous group velocity dispersion [30]. The ratio of the material's bandgap to the incident photon energy seems to determine the extent of anti-Stokes broadening [39].

It should be noted that almost all the earlier research work has been performed in the transparent window of the AGVD regime. In water there is significant linear absorption in the AGVD regime. There are only two reports on SC generation in the anomalous GVD regime in water, one using 1055 nm, 1 ps pulses in a 3 cm long cell [36] and the other using 1200 nm, 45 fs pulses in a water-filled photonic crystal fiber [40]. However, at these wavelengths water does not have significant absorption. Hence, it would be interesting to understand the effect of propagation in water at longer wavelength. This is important as wavelengths beyond 1300 nm are considered eye-safe wavelengths.

Filamentation in air at 3.1 μm (negative GVD regime) has been simulated by Shim et al. [41]. Propagation of ultrashort pulses in the anomalous group-velocity dispersion (GVD) regime in air has been theoretically investigated recently [41] but experimental studies of air filaments have been limited to the normal-GVD regime [42–47]. In case of air the wavelength corresponding to anomalous GVD occurs at 3.1 μm. With recent availability of laser technology, such as difference-frequency generation (DFG) [48] and optical parametric chirped-pulse amplification (OPCPA) [49, 50], pulses in the mid-infrared region with power >100 GW have become feasible. Availability of such pulses will allow experimental investigations of self-focusing in air in the anomalous-GVD regime.

Recently there has been renewed interest in generation of harmonics, particularly low-order harmonics from gases and air using ultrashort intense laser pulses. Such studies have profound implications in understanding basic mechanisms of filamentation nonlinear optics. Third harmonic generation (THG) using ultrashort pulses from air has been extensively studied; however these experiments have utilized 800 nm laser pulses [51–54]. In recent years, due to development of ultrafast sources in shortwave-infrared (IR) range, it has become possible to study higher harmonic generation and propagation in air [55, 56].

Both THG and optical Kerr effect (OKE) are related to the third-order susceptibility. A more generalised form of the Kerr effect, where the higher order terms are also considered, has been recently reported [57, 58]. The transient birefringence induced in different gases by intense ultrashort optical pulses has been interpreted as the saturation and sign reversal of the nonlinear refractive index prior to ionization of the medium, at an incident intensity of several tens of TW/cm^2. Ni et al. have reported measurement in atomic and molecular gases at low pressure and for short interaction length using long wavelength (1500 nm) 48 fs pulses [59]. Thus, apart from plasma, higher-order Kerr terms might provide an additional mechanism for defocusing and stabilising the femtosecond filaments [1, 60, 61]. There exist reports which support and contradict the claim [55, 59] that the higher-order Kerr effect (HOKE) provides a defocussing action. The assumption that intensity clamps in filaments suggests that intensities required for saturation of the Kerr effect cannot be reached in the filamentation process. In experiments with filamentation it is demonstrated that even if the higher-order Kerr terms exist, they are not operative in laser filaments in gases [62–64]. Recently, observation of a transition from plasma-driven to Kerr-driven laser filamentation was reported [65].

The validity of the HOKE model can be experimentally tested following the proposal by Kolesik and co-workers that involves a comparison of fifth harmonic (FH) and third harmonic (TH) efficiencies in gas media [66]. The ratio FH/TH is predicted to saturate at a high value of 0.1 in the HOKE model; in contrast the standard model (without HOKE) predicts this ratio to be very much lower (10^{-5}). It has been proposed that when the peak intensity exceeds turnover intensity (at which the Kerr effect would turn to zero, provided that HOKE is valid) the addition of higher-order Kerr terms affects the third (TH) and fifth (FH) harmonic efficiency [66]. Beyond the turnover intensity, the ratio of yields of the fifth to the third harmonic saturates [67]. In order to understand the difference in the harmonic yields in the two models let us consider how the fifth harmonic is generated. In the conventional approach, the fifth harmonic generation occurs via a cascaded process that involves nonlinear mixing of the already generated third harmonic and the un-depleted fundamental. On the other hand, in the new theory, the fifth harmonic is assumed to be generated directly from the fundamental through the higher-order Kerr response. It should be noted that higher-order harmonic cascading might occur from HOKE and the cascading contributions might modify the nonlinear index that is dependent on the wavelength [68].

In a recent report which tests HOKE model it has been shown that although the fifth-order nonlinearity is non-negligible, the overall defocusing effect via the higher-order nonlinearities is sufficiently small and it is the plasma formation that is a main defocusing mechanism in high power filamentation [69]. These workers have also explored the cross-phase modulation via the optical Kerr effect, and found that the higher-order nonlinearities can significantly alter the phase matching of harmonic generation.

In the following section we shall summarize some recent work on the propagation of long wavelength intense, ultrashort laser pulses through different media (solids, water and air). The spectral extent and spectral profile of the SC generated in transparent solids like BaF_2, CaF_2, fused silica and water in the normal and anomalous group velocity dispersion regime (AGVD) will be explored [70, 71]. The validity of the HOKE model will be tested under tight focusing condition [72]. A recent advance in application of such filamentation in biological media will be described. Specifically, we will discuss applications of propagation studies at longer wavelengths in DNA damage [73].

6.2 Propagation in Transparent Solids

Longer wavelengths correspond to the anomalous group velocity dispersion region for transparent solids like barium fluoride, calcium fluoride, and fused silica (see Table 6.1). Here, we discuss propagation of femtosecond pulses in transparent solids at wavelengths that correspond to normal, zero and anomalous GVD. Particularly, SC generation in these materials, and visualization of filamentation in BaF_2 along with estimation the 15-photon cross-sections, will be discussed.

In order to carry out experiments in all the three GVD regimes we require prior knowledge of GVD values for a given sample at specific wavelength. We estimate GVD values for various samples by adopting the following procedure. GVD is defined by the following equation:

$$GVD = \lambda^3/(2\pi c^2)\left(\frac{d^2n}{d\lambda^2}\right) \tag{6.1}$$

The dispersion is a material property and is deduced using the Sellmeier equation valid in the transparency region. The general form of this equation is [74]:

$$n^2(\lambda) = 1 + \frac{B_1\lambda^2}{\lambda^2 - C_1} + \frac{B_2\lambda^2}{\lambda^2 - C_2} + \frac{B_3\lambda^2}{\lambda^2 - C_3} + \cdots, \tag{6.2}$$

where the coefficients for BaF_2, CaF_2 and fused silica are obtained from a standard [75]. By solving the Sellmeier equation and substituting in the GVD equation we get the GVD values for specific wavelengths. In Table 6.1 we list the GVD values for specific wavelengths used in our measurements. In case of fused silica the wavelength at which the GVD value is zero is 1.27 μm, whereas for BaF_2 this wavelength is 1.92 μm and for CaF_2 it is 1.54 μm. We have also calculated the P_{cr} values at different wavelengths using the standard BGO model which are also listed in Table 6.1 [76].

6.2.1 Supercontinuum Generation

We used a Ti-sapphire laser (800 nm wavelength, 1 kHz repetition rate), with beam diameter 1 cm, 4 mJ energy, and 35 fs pulse duration in our experiments. The beam from this laser was used to pump an optical parametric amplifier (OPA) which generates wavelengths over the range 1.1–2.5 μm [70]. The residual 800 nm light was blocked using an RG850 filter. A pair of dielectric mirrors separated the signal and

Table 6.1 GVD and P_{cr} for different materials at three wavelengths

Wavelength (nm)	Fused Silica		BaF_2		CaF_2	
	GVD ($fs^2\ mm^{-1}$)	P_{cr} (MW)	GVD ($fs^2\ mm^{-1}$)	P_{cr} (MW)	GVD ($fs^2\ mm^{-1}$)	P_{cr} (MW)
800	36	2.5	37	3	27	4
1380	−10	7.4	16	9	6	12
2200	−107	19	−10	23	−33	30

idler wavelengths. The pulse duration of these IR beams was measured using a home-made autocorrelator. The pulse duration at 1.3 μm was ~56 fs and that at 2.2 μm was ~64 fs. The laser beam was focused using a 30 cm lens onto the samples such as fused silica, CaF_2 and BaF_2. Each sample was 15 mm in length. The supercontinuum (SC) was characterized using three different spectrometers. All our discussion pertains to only the axial part of the SC spectrum.

The spectral extent of the SC emerging from individual sample was measured using a combination of a lens and a spectrometer: three spectrometers were employed to cover the spectral range (200–2500 nm) of interest in our experiments. The incident laser energy was adjusted at each wavelength to ensure that all our measurements of the SC were made in the single filament regime and close to the threshold for SC generation. The SC conversion efficiency is an important parameter and was estimated by measuring the ratio of SC energy (in the range 400–1100 nm) to the incident energy.

In Fig. 6.1 we show the spectral profile of the SC as a function of incident wavelength. At 800 nm incident wavelength the GVD value for fused silica is positive: the extent of the SC covers the range 410–2080 nm, spanning more than 2 octaves. On increasing the incident wavelength to 1380 nm (GVD $= -10\ fs^2\ mm^{-1}$), the SC extent (410–2080 nm) was nearly the same as that for 800 nm. As the incident wavelength was further increased, to 2200 nm (GVD $= -107\ fs^2\ mm^{-1}$), there was an enhancement in the extension in the SC such that it covered the range up to 2300 nm. At this wavelength we observe a single isolated wing in the blue side with a peak at 530 nm. Earlier work in fused silica has shown an isolated wing in the visible part of the SC. This wing is observed to shift towards blue side as the incident wavelength increases [18, 30].

The SC conversion efficiency in case of 1380 nm incident wavelength was determined to be 25 %; it becomes 8 % for 2200 nm incident wavelength, both values being measured at an input energy of 160 μJ. Thus, our observations indicate that the SC efficiency reduces as the incident wavelength is increased [70].

SC generation was also explored in BaF_2 at longer wavelengths. In Fig. 6.2 we show the variation of the SC with incident wavelength. For both 800 and 1380 nm incident wavelength, the GVD value for BaF_2 is positive: the extent of the SC in case of 800 nm pump spans the range 320–1980 nm, while that for 1380 nm incident wavelength is 330–2240 nm, spanning more than 2 octaves. It may be noted that

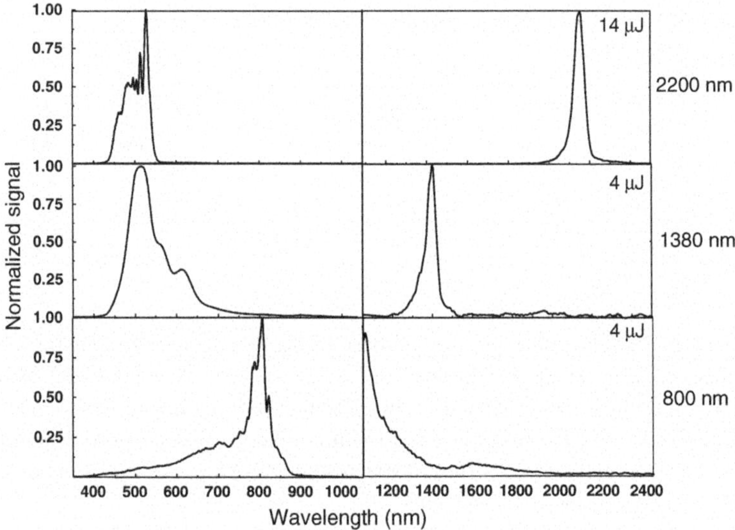

Fig. 6.1 Supercontinuum generation in fused silica with three different incident laser wavelengths: $40P_{cr}$ at 800 nm; $10P_{cr}$ at 1380 nm; and $10P_{cr}$ at 2200 nm. The *left* and *right panels* were measured using two different spectrometers. Note here that the spectra are normalised with respect to the peak wavelength in the respective spectrometers

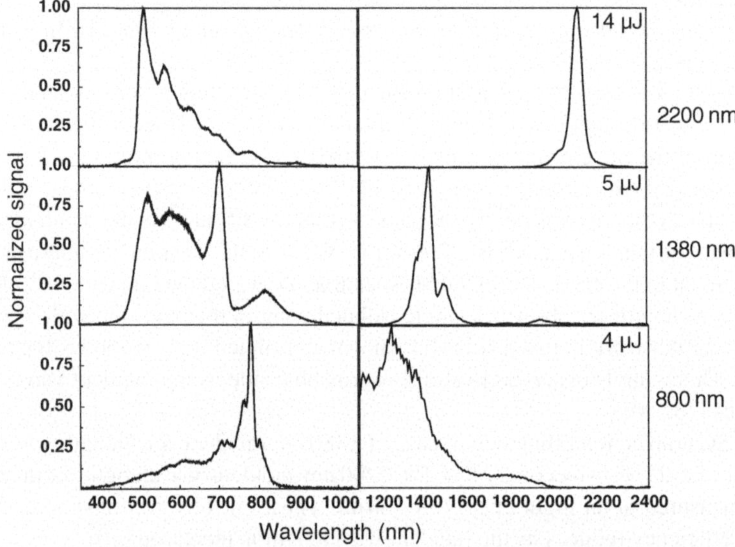

Fig. 6.2 Supercontinuum (SC) generation in BaF_2 with incident laser wavelengths: $33P_{cr}$ at 800 nm; $8P_{cr}$ at 1380 nm; and $10P_{cr}$ at 2200 nm. The *left* and *right panels* were measured using two different spectrometers. Note here that the spectra are normalised with respect to the peak wavelength in the respective spectrometers. Dip in SC from 750 to 900 nm is due to the presence of a highly reflecting dielectric mirror at 800 nm which allows us to observe the weak visible radiation

there is a dip in SC generation from 750 to 900 nm. This dip is due to the presence of a highly reflecting dielectric mirror at 800 nm. This allows us to observe the weak visible radiation. As the pump wavelength was increased to 2200 nm we access the negative GVD region (GVD is -10 fs^2 mm^{-1}) and we see a marked extension in the SC spectrum: the SC spectrum was observed to extend up to 2350 nm. The SC conversion efficiency was measured to be 32 % at 1380 nm incident wavelength and 13 % for 2200 nm incident wavelength, with both values being measured at an input energy of 160 μJ.

Figure 6.3 shows the SC spectra for CaF_2 as a function of different pump wavelengths. For both 800 and 1380 nm pump wavelengths the GVD value for CaF_2 is positive: the extent of the SC covers the range 300–2000 nm (in the case of 800 nm pump) and 320–2200 nm (for 1380 nm pump); both spectra span more than 2 octaves. Again, it is noted that there is a dip in SC generation from 750 to 900 nm, due to the presence of a highly reflecting dielectric mirror at 800 nm which allows us to observe the weak visible radiation. As the pump wavelength changes to 2200 nm, we access the negative GVD region (GVD is -33 fs^2 mm^{-1}) and we see a marked extension of the SC spectrum. It now covers the range 340–2450 nm. Note that even though the GVD value is -33 fs^2 mm^{-1} we still observe a shift in the SC emission towards shorter wavelength, similar to that in fused silica. However, the extent in the visible part of the spectrum is broader (370–850 nm) compared to that observed in

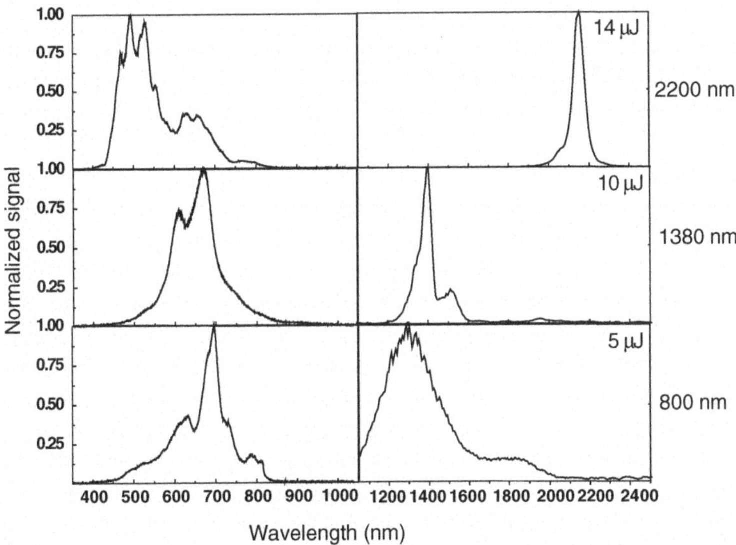

Fig. 6.3 Supercontinuum (SC) generation in CaF_2 at incident laser wavelengths: 30P_{cr} at 800 nm; 15P_{cr} at 1380 nm; and 7P_{cr} at 2200 nm. Note here that the spectra are normalised with respect to the peak wavelength in the respective spectrometers. Dip in SC from 750 to 900 nm is due to the presence of a highly reflecting dielectric mirror at 800 nm which was necessary to permit observation of the very weak visible radiation

Table 6.2 Variation of the anti-Stokes width with material properties

Bandgap (eV)	R	GVD $fs^2\ mm^{-1}$	Anti-Stokes width (nm)
7.5	8.3	−10	213
7.5	13.3	−107	104
9.1	16	−10	252
10.2	18	−33	284

R is the bandgap to photon energy ratio

fused silica (370–650 nm). The SC conversion efficiency was measured to be 32 % at 1380 nm incident wavelength and 14 % for 2200 nm incident wavelength, both at an input energy of 160 μJ. Recent work in CaF_2 has shown spectral broadening from 450 to 2500 nm in a 6 mm long crystal using 2 μm incident wavelength [19].

Note here that of all the three samples we have described, the extent of the SC spectrum is largest in the case of CaF_2. Moreover, this is accompanied by high conversion efficiency in the visible region at both 1380 and 2200 nm incident wavelengths. Midorikawa and coworkers [39] have shown that, in the normal dispersion regime, the ratio of material's bandgap energy to the incident photon energy is a parameter that determines the extent of anti-Stokes broadening.

In Table 6.2 we compare the ratio of the material's bandgap to the incident laser photon energy with the anti-Stokes width in the AGVD regime [70]. Our measurements show that for GVD values less than $-33\ fs^2\ mm^{-1}$, the anti-Stoke width increases with bandgap energy as well as with the ratio of bandgap to photon energy. Moreover, as the GVD value decreases (-10 to $-107\ fs^2\ mm^{-1}$) we observe a reduction in the anti-Stokes width in fused silica, in agreement with earlier measurements [18].

Earlier measurements in fused silica in the AGVD regime (at 1.5 μm) showed a broad maximum around 600 nm, with the spectral extent covering the range 400–950 nm [12, 17]. Filamentation in fused silica in the normal dispersion regime is observed to give rise to an extreme blue shifted continuum peak in the visible region, even though the filament is formed by near-IR pulses [38]. The blue-side peak has been identified as an axial component of the conical emission, as indicated by a three-wave mixing model [18]. Furthermore, anomalous dispersion in SC generation in fused silica has been shown to lead to formation of an isolated anti-Stokes wing (ASW) that is located in the visible region of the SC; this isolated ASW was described in terms of the interference of the SC light field encountering AGVD [30].

The highly asymmetric features of SC spectra may be due to odd-order dispersion terms that can be described in terms of spontaneous formation of stationary conical waves in a dispersive medium [27]. If the group velocity of the X-wave significantly differs from the group velocity of the incident laser pulse, one would expect a strongly blue-shifted peak in the pulse spectrum [38]. It can be also thought as scattering of the input pulse by the material waves constituting a nonlinear response that gives rise to SC generation [77]. Moreover, the dispersion properties of the medium become crucial ingredients in the SC formation process.

6.2.2 Filamentation

In order to study filamentation a BaF_2 crystal was utilized by us. This crystal enabled us to visualize the filaments due to multiphoton absorption induced fluorescence in the blue region, as described in earlier work [16]. A digital camera enabled us to capture the focusing-defocusing events within 15 mm long crystal. Another CCD camera coupled with an objective lens was positioned along the laser propagation direction to measure the size of filament. The band gap of BaF_2 is 9 eV. Irradiation by intense 800 nm laser light gives rise to six-photon-absorption-induced emission [13]. In our earlier measurements we have shown that the fluorescence peaks at 330 nm and extends towards 450 nm. Also, this blue fluorescence was a direct map of the intensity within the filament [16] and, thus, enables visualization of the propagating beam undergoing focusing-refocusing cycles when the power exceeds the critical power for self-focusing, P_{cr}. The number of cycles that are visualized depends on the peak power of the input pulse. Other effects, like diffraction, group velocity dispersion, self-phase modulation, and pulse self-steepening also contributes to the self-guiding process. Thus one can probe the effect of GVD by directly visualizing the filaments in the BaF_2 crystal at longer wavelengths. With 1380 nm wavelength incident light, due to 11-photon absorption, blue fluorescence is observed. At this wavelength, the GVD value is still positive.

In Fig. 6.4 the filament image shows the focusing-refocusing cycles at different values of incident laser power. There was no significant difference in the filament focusing-refocusing cycles compared to what we observed in our earlier measurements with 800 nm light. The choice of energy values in these measurements is such that only a single filament is visible in the propagation direction even though we observe focusing-refocusing events.

By changing the incident wavelength from 1.3 to 2.2 μm (Fig. 6.5) it was possible to access the anomalous GVD regime. In the case of 2.2 μm light, 16-photon absorption gave rise to the blue fluorescence as shown in Fig. 6.5. We observe focusing-refocusing cycles, with increase in incident energy from 26 to 90 μJ refocusing events are clearly observed to merge followed by extended fluorescence, the extended filament clearly demonstrates the role of GVD in the overall dynamics [70].

Earlier measurements in BK7 glass have shown that length of filament is larger in anomalous GVD (-25 $fs^2\,mm^{-1}$) regime compared to the case of normal GVD regime [22]. The separation between refocusing events was observed to be significantly larger in anomalous GVD regime. The mechanism responsible for these observations was attributed to energy transfer in the collapse region even after plasma arrest resulting in the formation of extended filaments. Theoretical and experimental work in AGVD regime on spectral transformation and spatiotemporal distribution of ultrashort laser pulses during filamentation in fused silica has been reported previously [21, 26, 78]. Practical realization of light bullets in AGVD regime has recently attracted a lot of attention [79–81]. The formation of quasi-solitons was first observed in a femtosecond laser pulse in the anomalous group velocity dispersion regime at

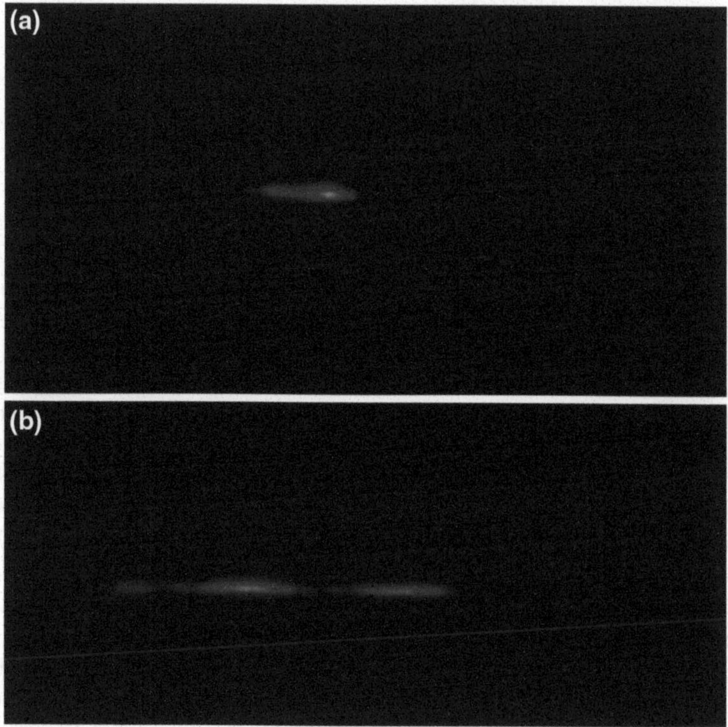

Fig. 6.4 Visualization of filamentation in BaF_2 at 1.3 μm (GVD value of 16 $fs^2\ mm^{-1}$) incident wavelength showing focusing-refocusing cycles at different values of incident laser power **a** 18 μJ ($35P_{cr}$), **b** 30 μJ ($60P_{cr}$). The filament lengths were measured to be **a** 0.2 cm and **b** 0.6 cm. The laser beam is incident from the *right side* of the images

a wavelength of 1900 nm with sub-GW excitation [79]. It was shown that the light bullets consists of a sharply localized high-intensity core that carries ~25 % of the total incident energy [80].

6.2.3 Estimation of Multiphoton Absorption Cross-Section in BaF_2

The filament in the transverse direction was imaged and the filament radius (L_{min}) inside the BaF_2 crystal at 2.0 μm wavelength was quantified. The filament radius was found to be 8.5 μm, a value that is almost double the value obtained at 800 nm [16]. We now discuss the estimates of peak intensities (I_{max}) and electron densities (n_e) by considering diffraction, the Kerr effect, and ionization responses. The estimates of the peak intensity, electron density, and the 15-photon absorption cross-section

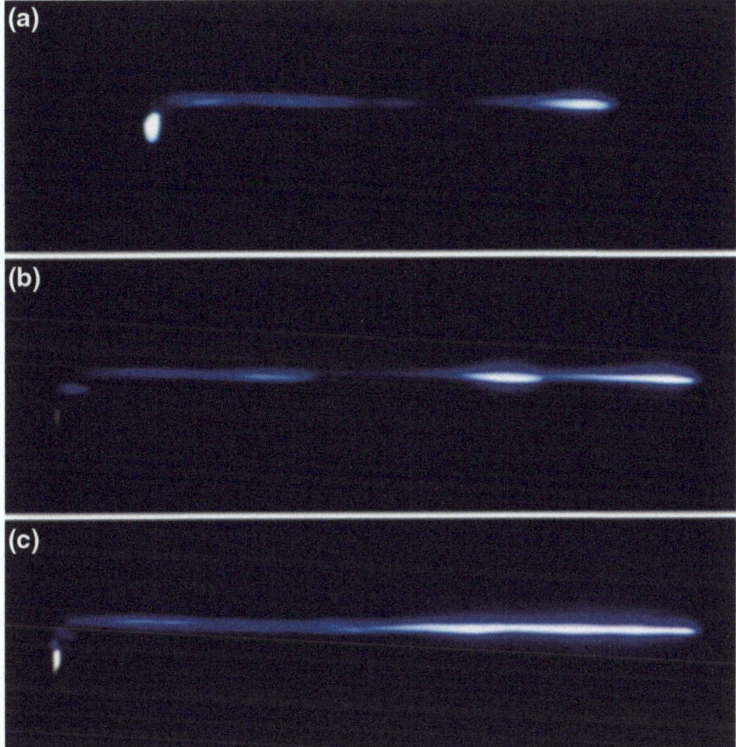

Fig. 6.5 Visualisation of filamentation in a 1.5 cm long BaF_2 crystal at 2.2 μm (GVD value of −10 $fs^2\,mm^{-1}$) with different incident energies **a** 26 μJ (18P_{cr}), **b** 45 μJ (30P_{cr}), **c** 90 μJ (60P_{cr}). The filament lengths were measured to be **a** 0.5 cm; **b** 0.75 cm; and **c** 1.0 cm. The laser beam is incident from the *right side* of the images. The *white spots* in the images on the *left hand side* are due to scattered white light

may be obtained from experimental observation of filament radius following the treatment described earlier [1, 2, 5, 16]. The estimated values of peak intensity (I_{max}) and electron density (ρ_{max}) within the crystal are 2.7×10^{13} W cm^{-2} and 1.7×10^{20} cm^{-3}, respectively. Using these values, the 15-photon absorption cross-section is estimated to be 6.5×10^{-190} cm^{30} W^{-15} s^{-1} [70].

6.3 Propagation in Water

Water has significant linear absorption in the anomalous dispersion regime. Below we show results of a systematic study of propagation in water at longer wavelength. Our measurements are performed over the wavelength range 800–1350 nm at which the GVD values are in the range from +24.8 fs^2 mm^{-1} to −100 fs^2 mm^{-1} [82]. In

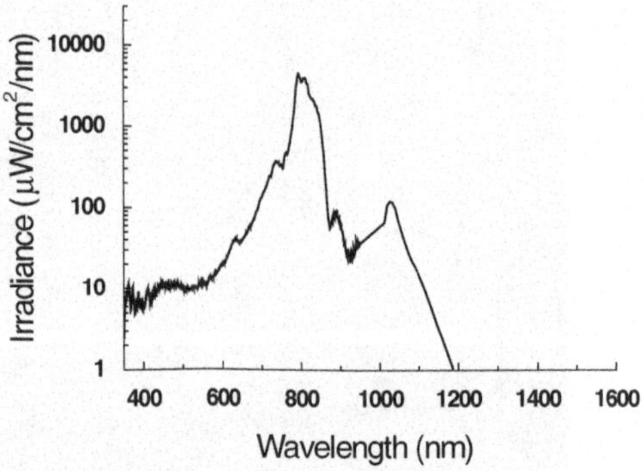

Fig. 6.6 Supercontinuum spectrum in water at 800 nm. GVD had a positive value of 24 $fs^2\ mm^{-1}$

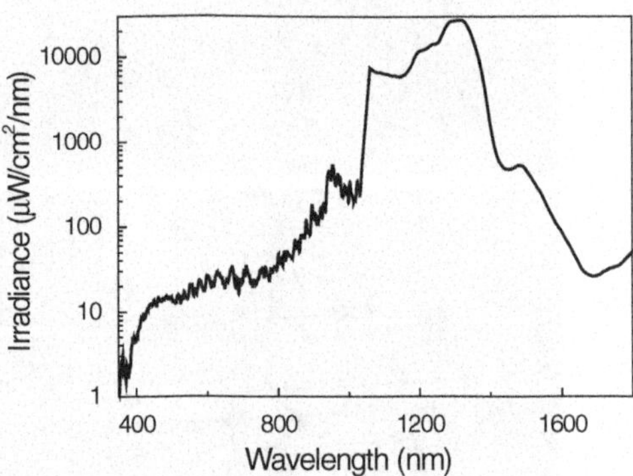

Fig. 6.7 Supercontinuum generation in water at 1250 nm incident wavelength. The incident energy was 3 μJ and GVD had a negative value of $-38\ fs^2\ mm^{-1}$

all the measurements that we discuss here the incident power was kept low enough to ensure that only a single filament was formed. A typical SC spectrum in water is shown in Fig. 6.6. We used 800 nm incident wavelength (GVD $= +24\ fs^2\ mm^{-1}$) where the linear absorption is very weak (0.02 cm^{-1}) [83]. In these measurements the laser energy was kept at 3 μJ (single filament regime). The spectrum was observed to extend over 350–1250 nm, corresponding to ~2 octaves.

The SC measurements carried out using 1250 nm wavelength light provide a negative GVD of $-38\ fs^2\ mm^{-1}$ (Fig. 6.7). At these wavelength water has linear absorption of 0.9 cm^{-1} and the SC is seen to extend from 350 to 1600 nm. At 1250 nm excitation, using higher laser power, we measured the efficiency of SC generation over the wavelength range 400–1100 nm to be 8 % [71].

On further increasing the incident wavelength to 1300 nm, where the linear absorption is significant (2.5 cm^{-1}), the energy required to form a single filament is found to increase to 5 μJ. The GVD at this wavelength is $-100\ fs^2\ mm^{-1}$. The extent of the SC ranges from 350 to 1600 nm, exceeding 2 octaves and is shown in Fig. 6.8a. However, the SC generation efficiency in this case is reduced to 6 %. We also note a prominent dip that appears near 760 nm, followed by a blue side continuum which peaks at 480 nm [71].

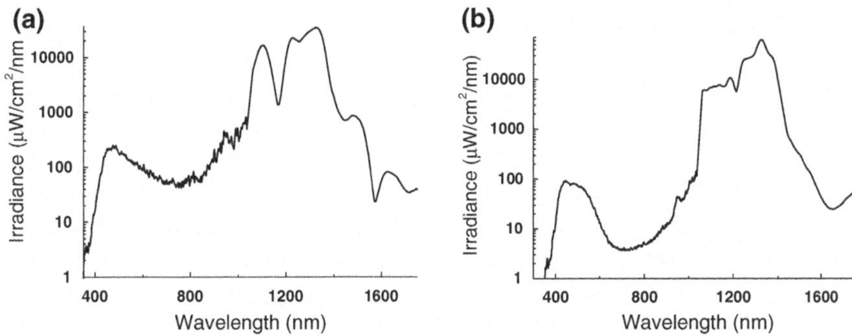

Fig. 6.8 Supercontinuum generation in water at incident wavelength of **a** 1300 nm (incident energy 5 μJ); **b** 1350 nm (incident energy 30 μJ)

Water has significant absorption (5 cm^{-1}) at 1350 nm, and negative GVD values in excess of $-100\,fs^2\,mm^{-1}$. The energy required to form a single filament in this case significantly increases to ~30 μJ and the SC (350–1600 nm range) extends more than two octaves. The blue shifted continuum is prominent as compared to 1300 nm; it peaks at 450 nm and is narrower compared to that obtained with 1300 nm incident light. The SC generation efficiency is reduced to 3 % and the SC spectrum has a prominent dip at 710 nm [71]. Similar narrowing and blue shifting has also been observed in the case of fused silica [18]. However, in the case of water, strong linear absorption occurs at pump wavelengths where anomalous GVD values can be accessed. Our results highlight the robustness of SC generation despite the presence of linear absorption in water. Thus linear absorption in water at this wavelength neither precludes the formation of the blue side continuum nor the extent of the SC.

To place our results in water in proper perspective, we consider earlier measurements in fused silica in the anomalous GVD regime at 1.5 μm central wavelength (with no significant absorption). A broad maximum around 600 nm was observed which extended from 950 to 400 nm [12, 17]. In recent work on a YAG crystal pumped at 3.1 μm, a structure was located in the SC spectrum near 610 nm [33]. Filamentation in fused silica in the anomalous dispersion regime is known to give rise to an extreme blue shifted continuum peak in the visible region, even when the filament is formed by near-IR pulses. The blue-side peak has been identified as an axial component of the conical emission, as indicated by a three-wave mixing model [18]. Furthermore, anomalous dispersion in SC generation in fused silica has been shown to lead to formation of an isolated anti-Stokes wing (ASW) that was located in the visible region of the SC; this isolated ASW was formed by the interference of the SC light field undergoing anomalous group velocity dispersion [30]. The blue side continuum peak that we observe in water at 480 nm may, therefore, be rationalized on the basis of two models: (i) we can consider it as an isolated ASW that is formed by the interference of the SC light field undergoing anomalous group velocity dispersion, as in the case of fused silica [30], or (ii) we can invoke a three-wave mixing process in which linear dispersion plays a crucial role; here the peak in the blue side would be ascribed to an axial component of the conical emission [77].

6.4 Propagation in Air

We now discuss experiments carried out in air to generate higher harmonics when using tightly focused ultrashort pulses of longer wavelengths (1.3 and 2 μm). The generated harmonics were reflected using a broadband dielectric mirror and were then focused on a spectrometer that was calibrated by using a standard calibration lamp source such that spectral irradiance could be determined. For measuring the harmonic efficiencies a power/energy meter along with appropriate interference filters were used. The incident energies used in our measurements varied from 50 to 300 μJ, much less than those used in another recent experiment [55], but, because of tighter focusing, we were able to reach comparable intensity values. At incident energy of 100 μJ (corresponding to an intensity of 41 TW/cm^2) the spectral irradiance ratio, fifth harmonic (FH)/third harmonic (TH), was measured to be 8×10^{-2}. Figure 6.9a shows the TH spectrum at an intensity of 73 TW cm^{-2} (200 μJ). The TH width in this case was measured to be 22 nm. The signal at 410 nm shown in Fig. 6.9b corresponds to FH generation from air. At these intensity values the TH energy is measured to be 8 $\pm$ 0.5 nJ, corresponding to an efficiency of 4×10^{-5}; FH efficiencies were less than the detection limit of our measurement system. In our experiments the coherence length is larger than the interaction length; hence harmonic generation will be phase matched but with a non-depleted pump condition. When the incident laser energy was 160 μJ (intensity of 65 TW cm^{-2}), a weak signal at 293 nm was observed, which corresponds to the seventh harmonic (SVH). At even higher energy the SVH signal was very distinctly observed, as is shown in Fig. 6.9c. At these intensity values the FH/TH spectral irradiance ratio is 12×10^{-2}, while the SVH/FH spectral irradiance ratio is $\sim 5.7 \times 10^{-3}$. The measurements performed at 280 μJ energy but using a shorter focal length lens of 3.5 cm correspond to an incident intensity of ~272 TW cm^{-2}, sufficiently high to observe TH, FH, and SVH. The TH energy at this intensity was 14 $\pm$ 0.8 nJ, corresponding to efficiency value of 5×10^{-5}. The FH energy was 4 $\pm$ 0.3 nJ, corresponding to efficiency value of $\sim 1.4 \times 10^{-5}$.

The FH/TH efficiency ratio was ~0.28 indicating that HOKE might be a dominant mechanism. The measurements performed using 1.3 μm incident wavelength show a similar trend for the FH/TH ratio. In Fig. 6.9d the FH/TH ratio as a function of incident intensity for both 1.3 and 2 μm wavelengths is shown. This data clearly indicates that in the case of 1.3 μm wavelength at higher intensity values the FH/TH ratio approaches saturation whereas no saturation of the FH/TH ratio is observed in case of 2 μm wavelength. The theoretically predicted [66] intensity at which saturation of the FH/TH ratio occurs is 50 TW cm^{-2}, much lower than the observed saturation intensity in our experiments (230 TW cm^{-2}). This may be attributed to different lengths of the interaction region considered in the theoretical prediction. In our experiments the incident power was not sufficient to attain the critical power for self-focusing (48 GW [83]) at 2 μm. Therefore, our efficiency values do not reach the 10^{-1} value reported in [66] considering HOKE. However, it is interesting to note

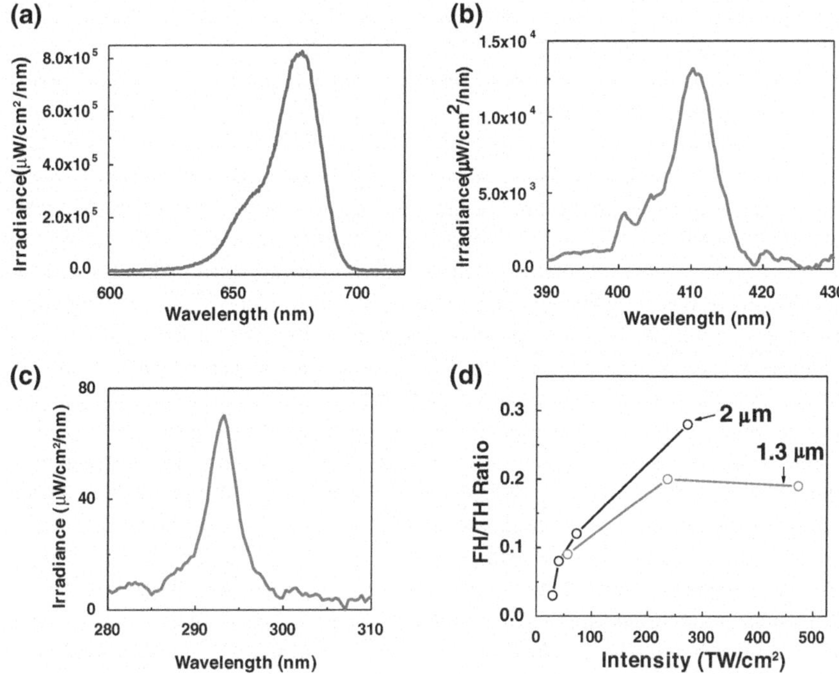

Fig. 6.9 Harmonic generation from air using 2 μm wavelength. **a** THG spectrum. **b** FH spectrum. **c** SVH spectrum. **d** Ratio of FH/TH as a function of incident intensity at 1.3 and 2 μm laser

here that in spite of such low TH conversion efficiency, we do observe FH and SVH at these power levels, and the FH/TH yield ratios are in agreement with those reported in [66] but at 1.3 μm incident wavelength.

6.5 An Application in the Life Sciences: DNA Damage

There are numerous applications of SC generation and filamentation, such as time-resolved broadband spectroscopy, generation of few cycle pulses and material modification [1, 84]. For illustrative purposes, we focus attention on a biologically relevant application using longer wavelength.

Femtosecond lasers find clinical application in dental surgery and eye surgery. [85]. It has been recently established that ultrashort pulses might induce damage to biomolecules such as DNA [73, 86–88]. Moreover, it was proposed that undesirable dose distribution inside tissue one of the major problems of radiotherapy might be circumvented by femtosecond filamentation [87]. This idea is quite remarkable since high power lasers particularly in infrared region can deliver high intensity pulses in a small volume that in principle surpass any clinical radiation source. Thus, one of the fundamental questions that arise naturally is what is the extent of damage that such pulses might induce in DNA?

During the propagation of intense femtosecond pulses through water multiphoton absorption and tunnel ionization occur and, hence, electrons are produced that are further accelerated by the electric field of the pulse by inverse Bremsstrahlung Plasma formation using ultrashort pulses in water has been well studied [89] and by treating water as an amorphous semiconductor with 6.5 eV band gap the breakdown process has been modeled. Nonlinear absorption in water is unique since it involves not only ionization but also dissociation, leading to formation of reactive radicals [89]. It has been shown earlier that an electron with $\sim$5 eV energy leads to formation of multiple transient anion states within DNA that decay into damaged structures involving single (SSB) and double strand breaks (DSB) [90]. Thus, ionizing properties of laser induced filamentation give rise to changes in the medium that has some resemblance to conventional high energy ionizing radiation, such as γ rays. It has been recently shown by Fricke dosimetry [91] that IR laser-pulse induced filamentation produces the same reactive species as the radiolysis of water by ionizing radiation. Earlier studies based on rate equations for optical breakdown in water indicate that electron densities of 10^{18}–10^{20} cm^{-3} can be attained [89] and these electrons may contribute to the formation of temporary negative ions in water + DNA. Subsequent breakup of such negative ions results in strand breakages. Strand break occurs when the sugar-phosphate backbone gets ionized when living matter is exposed to high-energy radiation. More than a decade ago, experiments on DNA in a thin film form established a new pattern showing that even electrons possessing a few eV of energy may induce strand breakages by formation of a dissociative temporary negative ion state [92, 93]. Therefore it is pertinent to carry out experiments on DNA in its native aqueous state.

Recent work using 45 fs pulses of intense (1–12 TW cm^{-2}) 820 nm light propagating through water in which DNA plasmids were suspended showed that initial supercoiled DNA transformed into relaxed DNA, as quantified by gel electrophoresis. The damage was caused by both slow electrons and OH radicals generated, in situ, during filamentation of femtosecond pulses in H_2O. A recent measurement that utilized intense light pulses of longer wavelength, 1350 and 2200 nm, at intensities in the TW cm^{-2} range demarcated the role played by OH radicals in inducing strand breaks in DNA close to physiological conditions [73]. The longer wavelength has an advantage since electrons do not directly induce DNA damage (Table 6.3).

It was found that not only a higher percentage of native supercoiled DNA structure becomes relaxed, compared to earlier work at 820 nm [86], but that there was also a tendency to form linear DNA. Linearization of DNA was a clear-cut signature of double strand breaks (DSB). It is important to note that such breaks are difficult to repair and were previously thought to be caused only by high-energy radiation. The measurements carried out with electron and OH scavenging chemicals indicate that at long wavelengths OH radicals induce both SSB and DSB. Thermal effects also give rise to SSBs, more with 2200 nm light than at 1350 nm. However, they have no role to play in inducing DSBs.

Table 6.3 Summary of experimental findings on DNA damage caused upon irradiation by intense ultrashort laser pulses

Wavelength (nm)	e-induced damage	OH-induced damage	Nature of damage	References
820	Yes	Yes	SSB	[86]
1350	No	Yes	SSB+DSB	[73]
2200	No	Yes	SSB+DSB	[73]

Damage occurs upon dissociative ionisation of the water medium, producing electrons and OH radicals. *SSB* single strand breakage; *DSB* double strand breakage

Laser beams of wavelengths $>1300\,\text{nm}$ are currently characterized as eye safe in industry and these findings that OH radical produced upon strong irradiation of water at such wavelengths can induce DSB in DNA have strong bearing on the safety concerning the eye [73].

6.6 Summary

In this chapter we discussed propagation of femtosecond pulses of longer wavelength in solid liquid and air particularly SC generation and filamentation in the AGVD region. A distinct isolated blue-side continuum was produced whose width narrows as the GVD value becomes more negative. In case of transparent solids such as fused silica, calcium fluoride and barium fluoride more than 2 octave-wide SC generation was observed. Extended filamentation was visualized in BaF_2 in the AGVD regime. By performing measurements in water at longer wavelengths we measure efficient SC generation despite significant linear absorption, highlighting the robustness of SC generation. Long wavelength propagation opens up new avenues for utility of multi-octave broadband sources.

Acknowledgments The Department of Science and Technology is thanked for assistance to JAD under the Women Scientists Scheme and to DM for the J C Bose National Fellowship. We also acknowledge the skills and continuing enthusiasm of colleagues A. Nath, K. Dota, R. Bernard, R. Deshpande, P. Vasa, M. Singh, R. Sharma, H. Bharambe, and J.S. D'Souza.

References

1. A. Couairon, A. Mysyrowicz, Phys. Rep. **441**, 47 (2007)
2. L. Bergé, S. Skupin, R. Nuter, J. Kasparian, J.P. Wolf, Rep. Prog. Phys. **70**, 1633 (2007)
3. R.R. Alfano, S.L. Shapiro, Phys. Rev. Lett. **24**, 584 (1970)
4. P.B. Corkum, C. Rolland, T. Srinivasan-Rao, Phys. Rev. Lett. **57**, 2268 (1986)
5. A. Braun, G. Korn, X. Liu, D. Du, J. Squier, G. Mourou, Opt. Lett. **20**, 73 (1995)
6. T. Brabec, F. Krauz, Rev. Mod. Phys **72**, 545 (2000)
7. S. Polyakov, F. Yoshino, G. Stegeman, J. Opt. Soc. Am. B **18**, 1891 (2001)

8. A. Dubietis, E. Gaizauskas, G. Tamosauskas, P. Di Trapani, Phys. Rev. Lett. **92**, 253903 (2004)
9. S. Skupin, R. Nuter, L. Bergé, Phys. Rev. A **74**, 043813 (2006)
10. A.K. Dharmadhikari, F.A. Rajgara, N.C.S. Reddy, A.S. Sandhu, D. Mathur, Opt. Express **12**, 695 (2004)
11. Y.H. Chen, S. Varma, T.M. Antonsen, H.M. Milchberg, Phys. Rev. Lett. **105**, 215005 (2010)
12. M.L. Naudeau, R.J. Law, T.S. Luk, T.R. Nelson, S.M. Cameron, Opt. Express **14**, 6194 (2006)
13. A.K. Dharmadhikari, F.A. Rajgara, D. Mathur, Appl. Phy. B **80**, 61 (2005)
14. A.K. Dharmadhikari, F.A. Rajgara, D. Mathur, Appl. Phys. B **82**, 575 (2006)
15. A.K. Dharmadhikari, K.M. Alti, J.A. Dharmadhikari, D. Mathur, Phys. Rev. A **76**, 033811 (2007)
16. A.K. Dharmadhikari, J.A. Dharmadhikari, D. Mathur, Appl. Phys. B **94**, 259 (2009)
17. A. Saliminia, S.L. Chin, R. Vallee, Opt. Express **13**, 5731 (2005)
18. M. Durand, K. Lim, V. Jukna, E. McKee, M. Baudelet, A. Houard, M. Richardson, A. Mysyrowicz, A. Couairon, Phys. Rev. A **87**, 043820 (2013)
19. J. Darginavicius, D. Majus, V. Jukna, N. Garejev, G. Valiulis, A. Couairon, A. Dubietis, Opt. Express **21**, 25210 (2013)
20. V.P. Kandidov, E.O. Smetanina, A.E. Dormidonov, V.O. Kompanets, S.V. Chekalin, JETP **113**, 422 (2011)
21. S.V. Chekalin, V.O. Kompanets, E.O. Smetanina, V.P. Kandidov, Quant. Electron. **43**, 326 (2013)
22. K.D. Moll, A.L. Gaeta, Opt. Lett. **29**, 995 (2004)
23. F. Wise, P. Di Trapani, Opt. Photonics News **13**, 28 (2002)
24. A. Houard, Y. Liu, A. Mysyrowicz, J. Phys: Conf. Ser. **497**, 012001 (2014)
25. J. Liu, R. Li, Z. Xu, Phys. Rev. A **74**, 043801 (2006)
26. L. Bergé, S. Skupin, Phys. Rev. E **71**, 065601 (2005)
27. D. Faccio, A. Averchi, A. Couairon, A. Dubietis, R. Piskarskas, A. Matijosius, F. Bragheri, M. Porras, A. Piskarskas, P. Di Trapani, Phys. Rev. E **74**, 047603 (2006)
28. A. Mysyrowicz, A. Couairon, U. Keller, New J. Phys. **10**, 025023 (2008)
29. M. Bradler, P. Baum, E. Riedle, Appl. Phys. B **97**, 561 (2009)
30. O. Smetanina, V.O. Kompanets, S.V. Chekalin, A.E. Dormidonov, V.P. Kandidov, Opt. Lett. **38**, 16 (2013)
31. O. Smetanina, V.O. Kompanets, S.V. Chekalin, V.P. Kandidov, Quantum Electron. **42**, 913 (2012)
32. O. Smetanina, V.O. Kompanets, S.V. Chekalin, V.P. Kandidov, Quantum Electron. **42**, 920 (2012)
33. F. Silva, D.R. Austin, A. Thai, M. Baudisch, M. Hemmer, D. Faccio, A. Couairon, J. Biegert Nat. Commun. **3**, 807 (2012)
34. A. Couairon, J. Biegert, C.P. Hauri, W. Kornelis, F.W. Helbing, U. Keller, A. Mysyrowicz, J. Mod. Opt. **53**, 75 (2006)
35. M. Hemmer, M. Baudisch, A. Thai, A. Couairon, J. Biegert, Opt. Express **21**, 28095 (2013)
36. M.A. Porras, A. Dubietis, E. Kucinskas, F. Bragheri, V. Degiorgio, A. Couairon, D. Faccio, P. Di Trapani, Opt. Lett. **30**, 3398 (2005)
37. M.A. Porras, A. Parola, P. Di Trapani, J. Opt. Soc. Am. B **22**, 1406 (2005)
38. D. Faccio, A. Averchi, A. Lotti, M. Kolesik, J.V. Moloney, A. Couairon, P. Di Trapani, Phys. Rev. A **78**, 033825 (2008)
39. A. Nagura, A. Suda, H. Kawano, M. Obara, K. Midorikawa, Appl. Opt. **41**, 3735 (2002)
40. J. Bethge, A. Husakou, F. Mitschke, F. Noack, U. Griebner, G. Steinmeyer, J. Herrmann, Opt. Express **18**, 6230 (2010)
41. B. Shim, S.E. Schrauth, A.L. Gaeta, Opt. Express **19**, 9118 (2011)
42. L. Bergé, S. Skupin, Phys. Rev. Lett. **100**, 113902 (2008)
43. I.G. Koprinkov, A. Suda, P. Wang, K. Midorikawa, Phys. Rev. Lett. **84**, 3847 (2000)
44. A.L. Gaeta, F.W. Wise, Phys. Rev. Lett. **87**, 229401 (2001)
45. L. Bergé, A. Couairon, Phys. Rev. Lett. **86**, 1003 (2001)
46. T.T. Xi, X. Lu, J. Zhang, Phys. Rev. Lett. **96**, 025003 (2006)

47. L.M. Kovachev, Opt. Express **15**, 10318 (2007)
48. P. Agostini, L.F. DiMauro, Contemp. Phys. **49**, 179 (2008)
49. O. Chalus, A. Thai, P.K. Bates, J. Biegert, Opt. Lett. **35**, 3204 (2010)
50. T. Popmintchev, M. Chen, P. Arpin, M. Gerrity, M. Seaberg, B. Zhang, D. Popmintchev, G. Andriukaitis, T. Balciunas, O.D. Mücke, A. Pugzlys, A. Baltuska, M. Murnane, H. Kapteyn, Science **336**, 1287 (2012)
51. H. Yang, J. Zhang, L.Z. Zhao, Y.J. Li, H. Teng, Y.T. Li, Z.H. Wang, Z.L. Chen, Z.Y. Wei, J.X. Ma, W. Yu, Z.M. Sheng, Phys. Rev. E **67**, 015401(R) (2003)
52. L. Bergé, S. Skupin, G. Méjean, J. Kasparian, J. Yu, S. Frey, E. Salmon, J.P. Wolf, Phys. Rev. E **71**, 016602 (2005)
53. R.A. Ganeev, M. Suzuki, M. Baba, H. Kuroda, I.A. Kulagin, Appl. Opt. **45**, 748 (2006)
54. S. Suntsov, D. Abdollahpour, D.G. Papazoglou, S. Tzortzakis, Phys. Rev. A **81**, 033817 (2010)
55. G.O. Ariunbold, P. Polynkin, J.V. Moloney, Opt. Express **20**, 1662 (2012)
56. D. Kartashov, S. Ališauskas, A. Pugžlys, A.A. Voronin, A.M. Zheltikov, A. Baltuška, Opt. Lett. **37**, 2268 (2012)
57. V. Loriot, E. Hertz, O. Faucher, B. Lavorel, Opt. Express **17**, 13429 (2009)
58. V. Loriot, E. Hertz, O. Faucher, B. Lavorel, Opt. Express **18**, 3011 (2010)
59. J. Ni, J. Yo, B. Zeng, W. Chu, G. Li, H. Zhang, C. Jing, S.L. Chin, Y. Cheng, Z. Xu, Phys. Rev. A **84**, 063846 (2011)
60. P. Béjot, J. Kasparian, S. Henin, V. Loriot, T. Vieillard, E. Hertz, O. Faucher, B. Lavorel, J.P. Wolf, Phys. Rev. Lett. **104**, 103903 (2010)
61. C. Brée, A. Demircan, G. Steinmeyer, Phys. Rev. Lett. **106**, 183902 (2011)
62. P. Polynkin, M. Kolesik, E.M. Wright, J.V. Moloney, Phys. Rev. Lett. **106**, 153902 (2011)
63. O. Kosareva, J.-F. Daigle, N. Panov, T. Wang, S. Hosseini, S. Yuan, G. Roy, V. Makarov, S.L. Chin, Opt. Lett. **36**, 1035 (2011)
64. Y.-H. Chen, S. Varma, T.M. Antonsen, H.M. Milchberg, Phys. Rev. Lett. **105**, 215005 (2010)
65. P. Béjot, E. Hertz, J. Kasparian, B. Lavorel, J.-P. Wolf, O. Faucher, Phys. Rev. Lett. **106**, 243902 (2011)
66. M. Kolesik, E.M. Wright, J.V. Moloney, Opt. Lett. **35**, 2550 (2010)
67. W. Ettoumi, P. Béjot, Y. Petit, V. Loriot, O. Faucher, B. Lavorel, J. Kasparian, J.P. Wolf, Phys. Rev. A **82**, 033826 (2010)
68. M. Bache, F. Eilenberger, S. Minardi, Opt. Lett. **37**, 4612 (2012)
69. D.L. Weerawarne, X. Gao, A.L. Gaeta, B. Shim, Phys. Rev. Lett. **114**, 093901 (2015)
70. J.A. Dharmadhikari, R.A. Deshpande, A. Nath, K. Dota, D. Mathur, A.K. Dharmadhikari, App. Phy. B **117**, 471 (2014)
71. P. Vasa, J.A. Dharmadhikari, A.K. Dharmadhikari, R. Sharma, M. Singh, D. Mathur, Phys. Rev. A **89**, 043834 (2014)
72. A. Nath, J.A. Dharmadhikari, A.K. Dharmadhikari, D. Mathur, Opt. Lett. **38**, 2560 (2013)
73. A.K. Dharmadhikari, H. Bharambe, J.A. Dharmadhikari, J.S. D'Souza, D. Mathur, Phys. Rev. Lett. **112**, 138105 (2014)
74. F.A. Jenkins, H.E. White, *Fundamentals of Optics*, 4th edn. (McGraw Hill, New York, 2001)
75. M.J. Weber, *Handbook of Optical Materials* (CRC Press, Roca Baton, 2003)
76. K. Dota, J.A. Dharmadhikari, D. Mathur, A.K. Dharmadhikari, Appl. Phys B. **107**, 703 (2012)
77. M. Kolesik, E.M. Wright, J.V. Moloney, Opt. Express **13**, 10729 (2005)
78. O. Smetanina, V.O. Kompanets, A.E. Dormidonov, S.V. Chekalin, V.P. Kandidov, Laser Phys. Lett. **10**, 105401 (2013)
79. M. Durand, A. Jarnac, A. Houard, Y. Liu, S. Grabielle, N. Forget, A. Dure'cu, A. Couairon, A. Mysyrowicz, Phys. Rev. Lett. **110**, 115003 (2013)
80. D. Majus, G. Tamošauskas, I. Gražulevičiūtė, N. Garejev, A. Lotti, A. Couairon, D. Faccio, A. Dubietis, Phys. Rev. Lett. **112**, 193901 (2014)
81. S.V. Chekalin, E.O. Smetanina, A.I. Spirkov, V.O. Kompanets, V.P. Kandidov, Quantum Electron. **44**, 577 (2014)
82. A.G. Van Engen, S.A. Diddams, T.S. Clement, Appl. Opts. **37**, 5679 (1998)
83. L. Bergé, Opt. Express **16**, 21529 (2008)

84. R. Li, X. Chen, J. Liu, Y. Zhu, X. Ge, H. Lu, L. Lin, Z. Xu, in Progress in Ultrafast Intense Laser Science I ed. K. Yamanouchi, S.L. Chin, P. Agostini, G. Ferrante (Springer, Berlin, 2006), pp. 259–273
85. S.H. Chung, E. Mazur, J. Biophotonics **10**, 557 (2009)
86. J.S. D'Souza, J.A. Dharmadhikari, A.K. Dharmadhikari, B.J. Rao, and D. Mathur Phys. Rev. Lett. **106**, 118101 (2011)
87. R. Meesat, H. Belmouaddine, J.-F. Allard, C. Tanguay-Renaud, R. Lemay, T. Brastaviceanu, L. Tremblay, B. Paquette, J.R. Wagner, J.-P. Jay-Gerin, M. Lepage, M.A. Huels, D. Houde, PNAS **109**, E2508 (2012)
88. D. Mathur, J. Phys. B:At. Mol. Opt. Phys. **48**, 022001 (2015)
89. A. Vogel, J. Noack, K. Nahen, D. Theisen, S. Busch, U. Parlitz, D.X. Hammer, G.D. Noojin, B.A. Rockwell, R. Birngruber, App. Phy. B **68**, 271 (1999)
90. L. Sanche, Eur. Phys. J. D **35**, 367 (2005)
91. R. Meesat, J. Allard, D. Houde, L. Tremblay, A. Khalil, J. Jay-Gerin, M. Lepage, J. of Phys: Conf. Series **250**, 012077 (2010)
92. B. Boudaiffa, P. Cloutier, D. Hunting, M.A. Huels, L. Sanche, Science **287**, 1658 (2000)
93. X. Pan, P. Cloutier, D. Hunting, L. Sanche, Phys. Rev. Lett. **90**, 208102 (2003)

Chapter 7
Dense Matter States Produced by Laser Pulses

Hiroaki Nishimura and Dimitri Batani

Abstract This paper addresses the study of ≪extreme≫ states of matter created by laser pulses. We define ≪Warm Dense States≫ and ≪High Energy Density≫, their importance in physics, how to produce and how to diagnose them, either using ns laser producing shock waves in matter or isochoric heating with short-pulse lasers.

7.1 Introduction

Recently ≪extreme states≫ of matter have become a new subject of study in science. These are defined as matter in extreme pressure conditions, either as warm dense matter (WDM) or high energy density (HED). Here we still provide an exact definition of WDM and HED and how to obtain them either by using shock waves produced by high-energy laser pulses with ns pulse duration, or by using high-intensity short-pulse lasers. We will also describe how to diagnose them in order to get the information concerning their density, temperature and equation of state.

WDM and HED are interesting from the point of view of basic physics since they describe either states, which are intermediate between cold solid and ideal plasmas, or dense plasmas which are characterized by strong coupling and partial degeneration. Also, HED and WDM are common to several fields of physics like astrophysics (and in particular planetology), inertial fusion (and in particular fast ignition and shock ignition research), and the physics of laser-driven radiation and particle sources.

We define WDM as matter at near solid density ($\rho \approx 0.01-100\,\mathrm{g/cm^3}$) and temperatures of a few eV ($1\,\mathrm{eV} = 11{,}400\,\mathrm{K}$).

H. Nishimura (✉)
ILE, University of Osaka, Osaka, Japan
e-mail: nishimu@ile.osaka-u.ac.jp

D. Batani
CEA, CNRS, CELIA (Centre Laser Intense at Applications),
University Bordeaux, UMR 5107, 33405 Talence, France
e-mail: batani@celia.u-bordeaux1.fr

K. Yamanouchi et al. (eds.), *Progress in Ultrafast Intense Laser Science XII*,
Springer Series in Chemical Physics 112, DOI 10.1007/978-3-319-23657-5_7

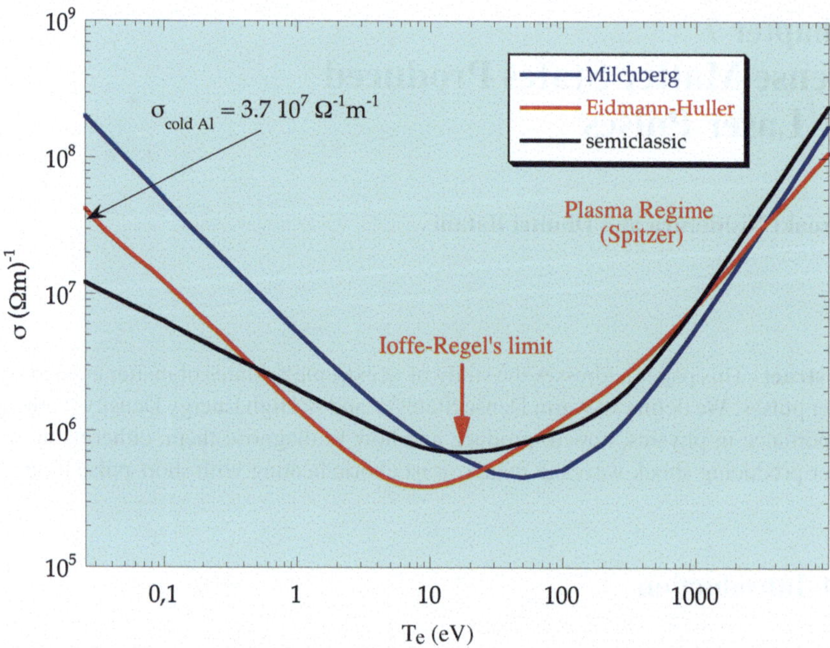

Fig. 7.1 Behavior of electrical conductivity of Aluminum at solid state density (2.7 g/cm^3) as a function of temperature. The graph shows the experimental data by Milchberg et al. [47] together with the results of a semi-classical calculation based on ionization degree as given by the SESAME tables, the results of a simplified Quantum model developed by Eidmann and Huller [48]. At $T_e \sim$ a few 10 eV (corresponding to the Fermi temperature of the material) a plateau is reached where collision rate is maximum (as given by Ioffe-Regel's limit). At larger temperatures matter enters in the plasma state described by the Spitzer's regime [49]

High energy density states are instead defined as matter with energy density $\geq 0.1\ \mathrm{MJ/cm^3}$ (corresponding to a pressure of 1 Mbar). These are conventional definitions but actually they are also related to ≪real≫ physical properties of materials.

For instance, concerning the definition of WDM, we can look at the electrical conductivity of Al in $(\Omega\,\mathrm{m})^{-1}$ versus T_e in eV at standard solid density, as shown in Fig. 7.1. We see that at low temperature electrical conductivity decreases when T_e increases. This is the normal behavior of solid density matter, which at low temperature is described by the usual quantum models of solid-state physics. This is due to the fact that by increasing material temperature the collision rate in matter increases, making electron motion more difficult. At large temperatures the behavior is reversed, conductivity increases with material temperature. This is the ≪plasma≫ regime which is described by Spitzer's theory. Plasmas are dominated by long-range Coulomb collisions, which become less and less effective as the kinetic energy of particles (hence the temperature) increases.

The intermediate range (plateau in the conductivity range) is situated at material temperatures of the order of the Fermi temperature of the material and corresponds to a collision rate, which is described by the so-called Ioffe-Regel's limit (i.e. the maximum possible collision rate: the electron mean free path becomes equal to the average ion-ion distance in the material). WDM describes the heating of solid density matter, the plateau and the transition to the plasma states. Hence WDM states can be defined as having $T_e \approx 1-100\,\text{eV}$ and $\rho \approx 0.01-100\,\text{g/cm}^3$ (to include all solid materials plus foams and aerogels).

Of course the behavior of electrical conductivity for insulators is different and even more complicated. At low temperature conductivity increases being driven by thermal ionization. At large temperatures again we enter in the Spitzer's regime. The actual calculation of the ionization degree (and hence of the number of free electrons) at a given density and temperature is a difficult problem which is closely connected to the description of equation of state. We assume thermal equilibrium and, in some cases, can make use of a Saha-like equation to calculate the density of the different ion charges and hence the mean ionization in matter. In any case, conductivity is monotonically increases over the while temperature range but still there is a change of regimes at a given temperature and density corresponding to the Ioffe-Regel's limit in conductors.

In all cases, either insulators or conductors, experimental data are very scarce. Also theoretical (numerical) calculations have practically only started in recent years. These are difficult because they just lie at the boundary between plasma physics and solid density physics. Matter is strongly coupled and degenerate (as in solids) but temperature is comparable or even larger than the typical temperature defining the material (i.e. the Fermi energy related to the density of free electrons in the material).

Coming now to ≪High energy density states≫, these are defined as matter with energy density $\geq 0.1\,\text{MJ/cm}^3$. In order to understand such definition we must recall that energy density is a pressure ($1\,\text{J/m}^3$ is equal to 1 Pa, and 10^5 Pascal is equal to 1 Bar. Therefore $1\,\text{Mbar} = 10^{11}$ Pascal $= 10^{11}\,\text{J/m}^3 = 10^5\,\text{J/cm}^3$). The typical pressure inside the material is related to the so-called ≪Bulk modulus≫ B of material, which is defined as

$$B = -V\frac{\partial P}{\partial V} \tag{7.1}$$

The bulk modulus B basically defines the ≪external≫ pressure which one must apply to "break" the bond in the material. For diamond $B = 4.42\,\text{Mbar}$. Then in order to reduce the volume of a diamond sample by 0.5 % we need a pressure $P \approx 4.42 \times 0.005 = 22\,\text{kbar}$. For most materials, $B \approx 1\,\text{Mbar}$, which explains why high energy density states are defined as matter with energy density $\geq 0.1\,\text{MJ/cm}^3$.

We can look at this issue from a complementary point of view. The binding energy, which relies one atom to the atomic lattice, is typically $E \approx 1\,\text{eV}$ per atom. The average inter-atomic distance in solids is typically ≈ 2 Å and therefore the energy density in solid matter is of the order of

$$E = \frac{1\,\text{eV}}{(0.2\,\text{nm})^3} = \frac{1.610^{-19}\,\text{J}}{810^{-24}\,\text{cm}^3} = 0.2 \times 10^5\,\text{J/cm}^3. \tag{7.2}$$

The big experimental problem connected to the study of WDM and HED is that it is difficult to diagnose such states of matter:

(1) Usually temperature is quite low, which implies a poor emission of radiation which in addition is centered at long wavelengths (VUV, Soft X-rays) which are difficult to collect (low transmission);
(2) Usually mass-density ρ is large. Hence it is difficult to use techniques like radiography, shadowgraphy or interferometric diagnostics (poor transmission of probe);
(3) Usually the produced sample is very small and the interesting state is transient. This implies the need for very high spatial resolution and very high temporal resolution.

Another problem is how to obtain WDM and HED states in the laboratory. Actually, this is in principle not difficult. For instance, as we will see later, in any laser ablation experiments (laser processing, laser machining) such states are produced. However it is much more difficult to obtain a sufficiently large sample, for a sufficient period of time, and with sufficient ≪good≫ properties (homogeneity and stationarity), to allow for a good measurement. Although there are several approaches used to create such extreme states described in the literature (explosions, including nuclear explosions, gas guns, X-FELs, Z-pinches, etc.), the two more reliable and more consolidated approaches rely on:

(1) Use of laser-driven shock waves;
(2) Use of short laser pulses to heat matter in the so-called ≪isochoric heating≫ approach. Such isochoric heating can be produced either directly by the laser or indirectly (e.g. using laser-produced X-rays, fast electrons, protons, ...).

These approaches are complementary. Shock waves allow getting densities bigger than that of the usual density of solids, but the reached temperatures are of the order a few 10 eV at most. Isochoric heating by definition allows only to reach solid-state density but in this case temperatures can be much higher, up to few 100 eV.

7.2 Production of Extreme States of Matter with Laser-Driven Shock Waves

In this approach, an intense laser is focused on the target surface producing laser-ablation of the material and plasma production. Such hot plasma (the plasma "corona") expands rapidly in vacuum and by conservation of momentum (Newton's action and reaction law) and intense shock is produced and travels inside the material.

The subject has long been studied starting in the 80s and scaling laws have been established to rely shock pressure to laser intensity and wavelength. In particular the most well-known scaling law is given by [1]:

$$P = 8.6\,(I/10^{14})^{2/3}\lambda^{-2/3}(A/2Z)^{1/3}\,\text{Mbar}, \tag{7.3}$$

where I is laser intensity in units of W/cm^2, λ is laser wavelength in units of μm, A is atomic mass number, and Z is atomic charge. Once the shock wave compresses the material to extreme states, we can try to measure its physical properties. In particular it is very interesting to obtain data on the equation of state (EOS) or materials in the Mbar and multi-Mbar pressure range. Presently, no models of material behavior for pressures above 10 Mbar have been experimentally validated and although in the limiting case of extremely high pressure, EOS are expected to be described by a Thomas-Fermi-Dirac model, the regimes of applicability and approach to this limit are not known.

One of the reasons is that experimental EOS data are sparse in this pressure range because of the difficulty in producing very high pressure and measuring the relevant parameters at the same time. Indeed while pressures <1 Mbar can be obtained with static methods (with a "diamond anvil cell"), higher pressures can only be obtained dynamically for short times by generating shock waves in the material.

In the past, only a few EOS measurements in the tens of Mbar domain could be performed by nuclear explosions. This, because of reproducibility, environmental, economic and proliferation considerations, was certainly not the best method of performing physics experiments and it is nowadays become impossible due to the Comprehensive Test Ban Treaty. Nowadays it is anyway possible to reach very high pressures in laboratory by using powerful-pulsed laser-generated shock waves, i.e. by focusing intense laser beams onto solid targets.

The study of EOS is particularly interesting for several fields of modern physics in particular astrophysics and inertial confinement fusion (ICF).

Usually astrophysicists deal with enough rarefied plasmas for which ideal gas physics is a good approximation. Unfortunately, however, this is not always the case. In particular in the center of stars and giant planets the density is so high that matter is ionized (called pressure ionization) and becomes a strongly coupled plasma that is the various ions interact strongly and no longer behave as free particles. Moreover, this is often accompanied by electron degeneracy as a result of the high electronic density.

Very high pressure can be reached with laser experiments, which can bring information on the interior of planets, giant planets, and stars. The theory of stellar evolution is affected by uncertainties in the equation of state in a few areas. One of this is white dwarfs. Through most of the white dwarfs, the pressure of degenerate electrons supports the material against gravity, but near the surface the electrons become less degenerate and the ions become important. The ions are also setting the specific heat and thus the rate at which the white dwarf can cool, a process that takes millions of years.

Another kind of dwarf stars, the brown dwarf, is poorly understood owing in part to our imperfect knowledge of the equation of state. Brown dwarfs have masses so low that they never got hot enough inside to burn hydrogen, but simply condense from the interstellar medium and shine dimly for a while as they get rid of the heat formed in condensing. Brown dwarfs have been observed experimentally in the last

20 years [2], and it's possible that they are so numerous that they comprise a large fraction of the mass in our galaxy (the mass which has been inferred to exist from the dynamics of the galaxy but has not been observed in luminous form). Their internal structure and cooling time depend on the details of the equation of state at densities approaching solid density at a temperature of a few eV. These conditions are easily reachable in laser experiments.

The other main field of physics interested to EOS is ICF. In ICF a thermonuclear target containing deuterium (D) and tritium (T) is illuminated by intense laser, X-rays or ion beams and implodes allowing the conditions for the nuclear fusion between D and T nuclei to be reached. Gigabar pressures are predicted in the spherically compressed capsules. The thermodynamics and hydrodynamics of these systems cannot be predicted without a knowledge of the EOS, which describe how a material reacts to extremely high pressure.

EOS experiments with shock waves are based on the so-called Hugoniot-Rankine relations, which, in the simplest case of "strong" shocks, read [3]:

$$\begin{aligned} \rho_0 D &= \rho(D-u), \\ P - P_0 &= \rho_0 D\, u, \\ E - E_0 &= u^2/2, \end{aligned} \tag{7.4}$$

where D, u, ρ, P and E are respectively the shock velocity, the fluid velocity, the mass density, the pressure and the specific internal energy of the material (after shock passage), while with "0" we indicate the values of the same quantities in the un-perturbed material (i.e. before shock passage). Being a system of 3 equations with 5 unknowns, the Hugoniot-Rankine relations allow one point on the EOS of the material to be found once two parameters are experimentally measured.

Now, although it has been well known for many years that lasers could produce shocks with pressures up to 100 Mbar [4], there was reluctance from the EOS community to use them as a quantitative tool in high-pressure physics.

This was due to the "poor" quality of produced shocks concerning the flatness of the shock fronts and constant shock velocities in the material. These requirements are essential to obtain accurate measurements of EOS, but another problem is even more important, that of preheating. Indeed hot electrons or hard X-rays can be produced in the plasma corona and penetrate into the material ahead of the shock and "preheat" it. In this case, the shock wave travels in a non-unperturbed material and the Hugoniot-Rankine relations are useless.

The possibility of creating spatially very uniform shocks in solids with negligible preheating was demonstrated in the 90s by producing shock waves by direct laser drive with optically smoothed laser beams [5].

Direct drive experiments were mainly affected by the uniformity problem, which is connected to the laser intensity distribution whole over the focal spot (not flat but, in first approximation, gaussian and with many speckles or hot spots). Thanks to the introduction of optical smoothing techniques like random phase plates (RPP) [6] and phase zone plates (PZP) [7] (arrays of Fresnel, each with a random dephasing of 0 or

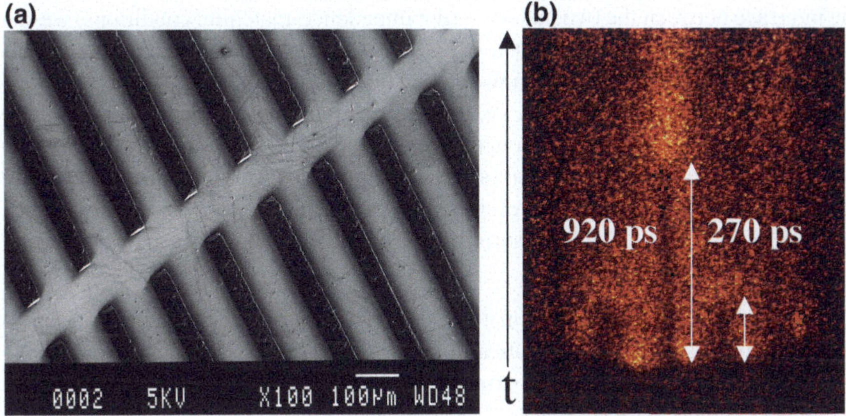

Fig. 7.2 **a** SEM image of carbon steps with $\rho_0 = 1.45\,\text{g/cm}^3$ deposited on a CH/Al substrate. Al steps are not present since they were deposited later. **b** Shock breakout streak image of the target rear side in emission. Shot energy was 25.3 J. *Arrows* indicate the shock breakout from the Al step (*right*) and from the C step (*left*). The size of the image is $600\,\mu\text{m} \times 1.7\,\text{ns}$

π to break the laser beam spatial coherence and give smoothing effects) high quality flat shock waves were generated. This allowed high pressures (10–50 Mbar) to be reached over a sufficiently large spot with relatively small lasers (=100 J).

Then one needs to diagnose the shock. The simplest diagnostic is based on the shock emergence from the target rear face (shock chronometry). It is based on the detection of the emissivity, in the visible region, of the target rear face, illuminated on the other side by the laser beam. The arrival of the shock causes a sudden variation in temperature and hence of emissivity. A photographic objective images the rear surface onto the slit of a visible streak camera in order to record such variations in time. Using a stepped target, shock velocity can determined with high precision by measuring the shock emergence time from the base and from the step of the target. The method is shown in Fig. 7.2.

Finally, in order to obtain an EOS point one must solve the Hugoniot-Rankine systems of equation. This can be done by measuring two parameters for the same material in the same laser shot. This approach was for instance used at Lawrence Livermore National Laboratory to obtain the EOS data for deuterium [8]. Here they also measured the fluid velocity through time resolved X-ray radiography. Alternatively, instead of measuring 2 parameters to determine the EOS, we can use the "impedance-matching technique" [3] with two-step, two-material targets. Here a "relative" EOS point of one "unknown" material can be obtained if the EOS of a "reference" material is assumed to be well known. In many cases, the target is made of a base of aluminum (chosen as reference material), which supports two steps, one of aluminum and the other one of the material to be investigated. This target geometry allows the shock velocities to be experimentally determined in the two materials on the same laser shot. By knowing the aluminum EOS and using the laws of shock transmission at the interface between the two materials (the impedance-matching conditions) we can then find the unknown EOS points.

The methods described until now have a limitation that they only allow to measure so-called "Hugoniot states", i.e. the points lying on the Hugoniot curve of the material, those which, by definition, can be obtained with single shock compression starting from the material in usual (standard) conditions. But many more states are present in the phase plane! Therefore it is very interesting to design experiments, which allows producing and measuring "off-Hugoiot" states. This can be done using different approaches:

(1) Begin with a non standard density, e.g. using porous materials (foams, aerogels)
(2) Precompress the material statically (this allows getting the same compression with lower temperature, in particular one can then get closer to the isentrope curve of giant planets)
(3) Precompress the material using multiple shocks or using laser pulse shaping to get closer to isentropic compression (a double shock technique has been sued and is particularly simple)
(4) Use release states, i.e. allows partial expansion of the compressed material before measuring its properties.

7.3 An Example: High-Pressure Equation of State of Carbon

In the Megabar pressure range, extreme states of matter are reached and many materials undergo interesting phase transitions. For instance, materials like hydrogen and carbon are expected to become metals. The knowledge of such states of matter, and in particular of the equation of states of materials, are of extreme importance in astrophysics and planetology since pressures of several Megabars or tens Megabars are indeed reached in the interior of planets, giant planets, and brown dwarfs.

One particularly interesting material is carbon. Its equation of state (EOS) at high pressures (Megabar or Multi-Megabar regime) is of interest for several branches of physics, namely:

Material Science

Carbon is a unique element due to its polymorphism and the complexity and variety of its state phases. The EOS of carbon has been the subject of several recent important experimental and theoretical scientific works [9–17]. The important phenomenon of carbon metallization at high pressure has long been predicted theoretically but until now never experimentally proved. At very high pressures the regime of non-ideal strongly-correlated and partially-degenerate plasmas is approached, which is characterized by an almost complete absence of experimental data [18–20].

Astrophysics

Description of high pressures phases is essential for developing realistic models of planets and stars data [21, 22]. Carbon is a major constituent (through methane and carbon dioxide) of giant planets like Uranus and Neptune. Very large magnetic

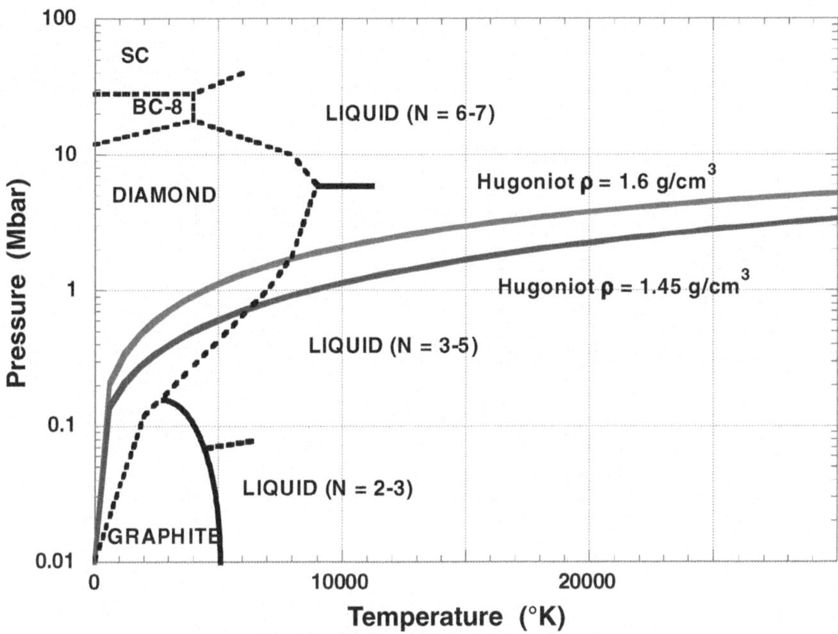

Fig. 7.3 Grumbach and Martin's phase diagram [13] and the two Hugoniot curves corresponding to the initial densities $\rho_0 = 1.6\,\mathrm{g/cm^3}$ and $\rho_0 = 1.45\,\mathrm{g/cm^3}$

fields have been measured by Voyager 2 in planets like Uranus and Neptune. Such intense, asymmetrical magnetic field is supposed to be originated in the mantles of these planets, the so called ≪hot ices≫ made of H_2O, NH_3, and CH_4. Therefore the presence of magnetic fields requires the existence of a fluid, conducting region (based on water, ammonia or carbon) to produce them by dynamo effect [23, 24]. High pressures could produce methane pyrolysis with a separation of the carbon phase and possible formation of a diamond or metallic layer. Metallization of the carbon layer could then provide the needed high electrical conductivity.

In Fig. 7.3 we have reported a simplified version of Grumbach and Martin's phase diagram [13] to which we added the Hugoniot curves corresponding to the initial densities $\rho_0 = 1.6\,\mathrm{g/cm^3}$ and $\rho_0 = 1.45\,\mathrm{g/cm^3}$ (the two values used in one of our experiments), calculated following the Sesame tables [18]. Again, the liquid metallic phases can be easily reached with laser shocks.

First Hugoniot data for carbon with laser-driven shocks have been obtained in [25] in the pressure range 1–8 Mbar, as shown in Fig. 7.4. In this work, a total of 9 new EOS points was obtained against a total of about 20 points which were at the time available in literature for $P > 1$ Mbar [26–30]. This shows another advantage of laser experiments, namely the fact that they allow increasing the statistics in the HED regime.

The experiment was based on generating high quality shocks and using "two steps-two materials" targets (Fig. 7.2). Relative EOS data of "unknown" materials

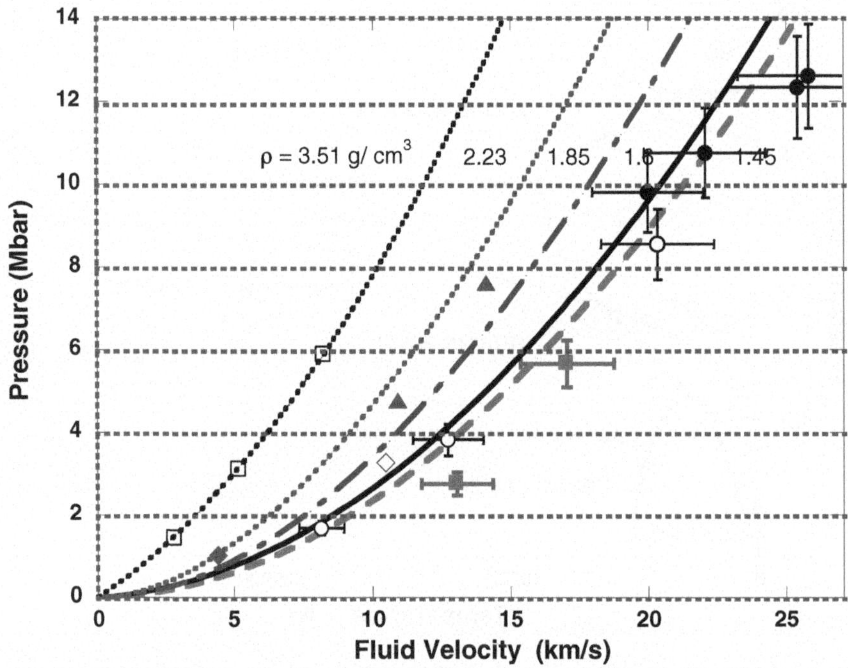

Fig. 7.4 Experimental EOS results from shock experiments. Only data with $P \geq 1.5$ Mbar and corresponding Hugoniot are shown. Points from [25]: *full squares* 1.45 g/cm^3(obtained at LULI laboratory); *empty circles* 1.6 g/cm^3 (LULI); *full circles* 1.6 g/cm^3 (obtained at PALS laboratory). Previous points: *empty diamond* 1.85 g/cm^3 Pavlovskii et al. [26]; *triangles* 2.2 g/cm^3 Nellis [28]; *full diamond* 2.23 g/cm^3 Pavlovskii et al. [26]; *empty squares* 3.51 g/cm^3 (*diamond*) Pavlovskii [27]

(here C) are obtained by using a "well-known" reference (here Al). Al behavior at high pressure is well known, making it a typical reference material for shock experiments. Figure 7.2 also shows a scanning electron microscope (SEM) photo of the carbon steps deposited on a CH/Al substrate. The deposition technique allowed the realization of targets with an acceptable surface roughness (less than 0.5 μ, i.e. $\approx$ 3 % of step thickness which was of the order of 15 μm). These give an error comparable to the typical $\leq$5 % due to streak camera resolution. The Al step thickness was 5 μm.

7.4 Isochoric Heating

Matter can be taken to WDM states by fast delivery of energy. Let's consider as an example a target made of an Al foil of size 200 μm $\times$ 200 μm $\times$ 10 μm. Since the density of Al is 2.7 g/cm^3, this corresponds to a volume $V \approx 4 \times 10^{-7}$ cm^{-3} and a mass of 1 μg, or 4×10^{-8} moli, i.e. it contains 2×10^{16} atoms. Delivering 1 J of energy (1 eV $= 1.6 \times 10^{-19}$ J) would imply an average energy per atom of the order of $\approx$300 eV. In reality, however, the calculation should be more complex: Al is certainly ionized and this increases the number of "particles", which share the energy. For example if $Z^* \approx 9$ we get $T_e \approx 30$ eV. Notice that the energy per unit volume is $1\,\mathrm{J}/4 \times 10^{-7}\,\mathrm{cm}^{-3} = 2.5\,\mathrm{MJ/cm}^3$, well within HED regime.

However, in order to reach such state and get "isochoric heating" the deposition of energy must be very fast. This is due to several reasons: First, let's consider that at $T_e \approx 30$ eV the sound velocity is in the plasma is of the order of

$$c_s = 9.8 \times 10^5 \sqrt{Z^* T_e / M}\ \mathrm{cm/s} \approx 3 \times 10^6\ \mathrm{cm/s}. \tag{7.5}$$

The rarefaction waves proceeds at c_s and then, over a time t = 15 ns, the target expands of $\pm 50\,\mu$m and therefore its volume becomes 10 times larger. The initial electron density in Al is $n_e \approx 6 \times 10^{22}\,\mathrm{cm}^{-3}$ drops to 1/10. In general terms, this means that if expansion is too large, the HED regime will not be reached. If density is high but changes with respect to solid state, then we are still in the HED regime but we need additional diagnostics to measure it. Therefore, rapid delivery of energy must take place over a time, which is much shorter than hydro timescale in order to maintain the density.

Then two approaches are possible with respect to target design:

(1) Tampered targets: a buffer material constraints expansion and keeps the density of an intermediate layer of the material to be studied. The problems of this approach is that the buffer material also absorbs the energy deposited by the laser.
(2) Small size targets (at limit: use of nm size clusters). In this case the energy is completely "used" in the material to be studied but the problem is that it is difficult to control the hydrodynamics due to "prepulse" (in particular, the smaller the target the more sensitive it is to expansion problems).

Also it is very important that the deliver of the energy must take place over a distance comparable to the target thickness in order to get a rather uniform temperature, which implies a homogenous sample easier to be diagnosed. Laser irradiation cannot penetrate deep into the target. A preplasma is always formed in front of the solid material and the laser can penetrate only up to the critical density (of the order of 10^{21} electrons/cm^3). Therefore the laser cannot be directly used to heat the material in depth. Again, different approaches have been used and described in the literature:

- Fast (relativistic) electrons created by ultra high intensity (UHI) laser-matter interaction [31]
- Cluster irradiation [32]
- Tampered targets with X-rays (produced from laser-plasmas) [33]
- Proton beams from UHI interaction [34]

Finally, in order to achieve isochoric heating with fast electrons, it is important that the sample is homogeneously heated, i.e. no strong temperature gradients are created inside matter (otherwise one would only get an "average" measurement). For instance, fast electrons created in the interaction of ultra-high-intensity lasers with solid targets, can propagate in the target and deposit their energy thereby heating it in depth. Also, when a thin target (thickness d) is used:

(1) Fast electrons cannot escape because they are effectively confined in the target by the huge electrostatic fields that are created by charge separation as soon the very first fast electrons leave the target.
(2) Since target thickness is much smaller than the typical propagation range of fast electrons in matter ($d \ll R$), the electrons reflux many times in the target before losing their energy.
(3) This takes place on a very fast time scale $t \approx d/c = (0.01\,\text{cm})/(3 \times 10^{10}\,\text{cm/s}) = 3 \times 10^{-13}\,\text{s} = 0.3\,\text{ps}$. Therefore matter is considered to be uniformly heated in a very short time scale producing isochoric heating.

7.5 Creation of WDM with Hot Electrons

A giant knot of energetic electrons seems a good energy carrier for WDM creation. Due to interaction of an intense laser pulse with plasma formed on a material surface, very energetic electrons called hot electrons are generated with a nearly Maxwellian spectrum, with temperature [35]

$$T_{\text{hot}} = \left[\left(1 + \frac{I_L \lambda_{\mu\text{m}}^2}{1.37 \times 10^{18}}\right)^{1/2} - 1\right] m_0 c^2, \tag{7.6}$$

where I_L is laser intensity in units of W/cm^2, λ is laser wavelength in units of μm, m_0 is electron static mass and c is light velocity. For example, T_{hot} becomes 0.161 MeV at $I_L = 10^{18}\,\text{W/cm}^2$ for $\lambda_{\mu\text{m}} = 1$, and electrons of this energy penetrate in a range R of 483 μm for a plastic (density ρ of $1.0\,\text{g/cm}^3$) using a formula of $R = 0.6 f_R T_{\text{hot}}$ (g/cm^2) assuming $f_\text{R} = 0.5$ [36]. In case of laser produced plasma, however, this penetration range is not simply applied because forward current, caused by the rapidly traveling electrons, is well above the Alfven current limit $I_\text{A} = 17\beta\gamma$ kA where β and γ are the relativistic β and γ factors [37]. In the case of $I_\text{L} = 10^{18}$ W/cm^2, the electric current carried by the absorbed electrons becomes 10 MA for a 20-μm-diameter beam whereas the Alfven limit current is 22 kA. Therefore, a magnetic field induced by the forward electron stream becomes large enough that Larmor radius of the electrons becomes smaller than the beam radius. As a consequence, magneto hydrodynamic instabilities are induced and the beam electrons are not further transported simply in the traveling direction. Meanwhile bulk electrons in a matter, having higher density and lower energy, flow as a counter stream of the hot electrons, and will play an important role in the energy deposition process [38]. In this way, hot electrons seem a very good candidate for WDM creation in terms of their energy density and range but a careful treatment is needed in the study of energy transport and deposition by hot electrons.

7.6 Creation of WDM with Radiation

Laser produced high-Z matter emanates intense thermal radiation that can also be a good source for WDM creation. Depending on plasma temperature and its constituents, radiation spectrum ranges from 0.1 to 10 keV. In particular if radiation plasma is optically thick, the spectrum fits with a Planck distribution [39]. Radiation temperature becomes as high as 500 eV, corresponding to $2.5 \times 10^{14}\,\mathrm{W/cm^2}$ at the radiator surface.

A good example of radiative burnthrough [40] with laser-produced plasma is shown in Fig. 7.5. A 0.5-μm-thick beryllium foil was heated by thermal x rays from a laser-produced Au plasma. Time resolved absorption spectra for the photoionizing process were observed in the spectrum region of 100–250 eV. Blue shift of absorption K-edge due to temporal evolution of ionization stage is clearly observed. The experimental results were compared with radiation hydrodynamic simulations

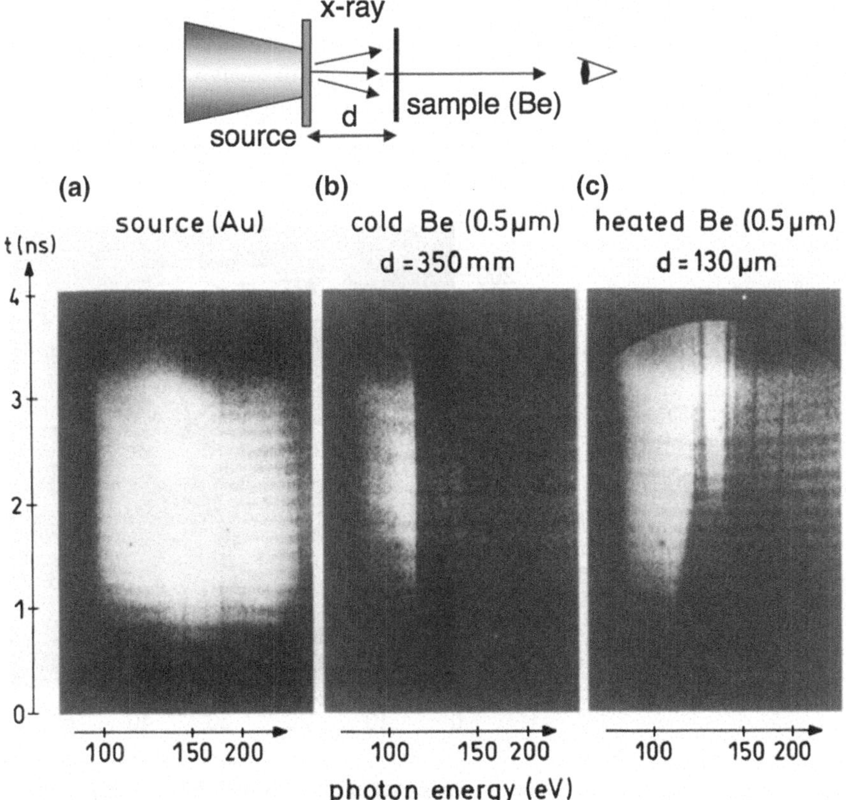

Fig. 7.5 Time-resolved spectrum obtained **a** with the source alone, **b** with a cold, and **c** with an x-ray heated 0.5-μm-thick beryllium foil [40]

post-processed with a detailed atomic kinetic model. A good agreement was obtained in the comparison, inferring that Be was heated up to temperatures of a few 10's eV. Smoothing of the Be K edge observed was attributed to thermally smoothed Fermi edge.

In general, x-ray pulse generated with an ultra-short, high intensity laser is so short that thermal equilibrium state is hardly established. Instead, inner-shell ionization dominates over ionization process creating so-called hollow atoms [41]. This matter state is important to study detailed energy relaxation processes in a WDM.

7.7 Physics of Isochoric Heating with Laser-Driven Hot Electrons

As discussed in the above section, hot electrons generated by high intensity laser pulse are very susceptive to local self-induced electromagnetic fields. Typical behavior is refluxing of hot electrons around a thin or reduced-mass material as shown in Fig. 7.6 [42]. Electrons are bounded within the target volume by the large electric fields formed at any surfaces. This behavior enhances homogeneity of material heating and x-ray yield arising from inner-shell ionization by electron impact but reduces energy localization, then temperatures. Thus, isochoric heating (heating of material

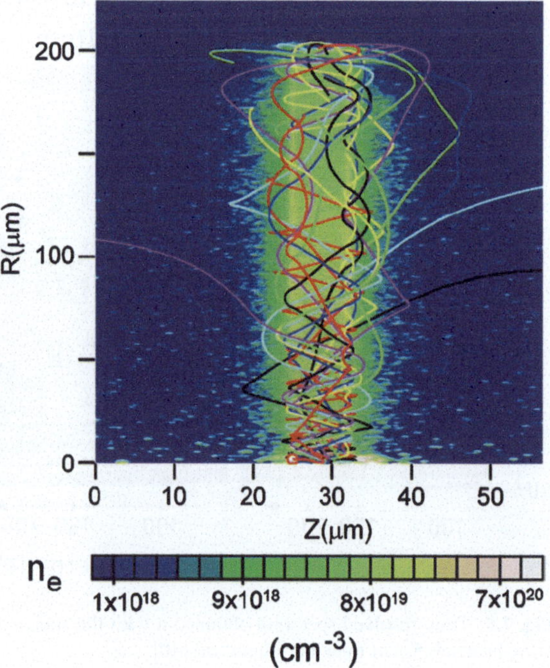

Fig. 7.6 Particle simulation of thin foil target irradiated with an intense laser pulse at the *left side*. The *color* represents the hot electron density and the *lines* represent electron trajectories. Back-and-forth motion between the *front* and *rear sides* of target is seen: electron refluxing [42]

without volume change due to disassembly) of dense matter was investigated by using a reduced mass target irradiated with an ultra-short high intensity laser pulse [43, 44].

Three heating processes occur in laser-irradiated dense matter [38]. One is a collisional energy transfer directly from hot electrons to bulk electrons called drag heating. The second one is a scattering of cold bulk electrons, flowing as a counter stream of hot electrons, by ions, called resistive heating. The third one is a diffusive thermal transport among bulk electrons having a sharp gradient in temperature locally. After the onset of refluxing, volumetric, uniform heating can be expected as far as hot electrons recirculating everywhere inside target by the drag and/or resistive heating. In the following, physics of material heating with laser-driven hot electrons will be discussed in detail by citing the study of [43].

In the following, we present result of a research with, a triple-layered target consisting of parylene (C_8H_8) and polyvinyl chloride (PVC; C_2H_3Cl) were used. Chlorine, a lower-Z material, was chosen specifically because it is widely adopted as a spectroscopic tracer material for fusion capsules made of plastic. Average ionization Z values are 3.5 and 5.3 respectively for fully ionized parylene and PVC, so material mismatching at the interface is expected to be less important. Square shape targets were fabricated with different side length L from 50, 100, 300 to 1000 μm in order to control the total mass of the target.

The target was irradiated with an ultra-short laser pulse at 10^{18} W/cm^2 and x-ray emission spectra from the PVC tracer layer were observed to extract bulk electron temperatures. Kα lines from partially ionized chlorine embedded in the middle of a triple-layered plastic target were measured to evaluate bulk electron temperature in the tracer region inside the target. Typical x-ray spectra from the conical crystal spectrometer for five different types of targets are shown in Fig. 7.7a–e respectively for (a) $L = 50\,\mu$m, (b) $L = 100\,\mu$m, (c) $L = 300\,\mu$m, (d) $L = 1000\,\mu$m, and (e) the double layered target without the surface overcoat. These spectra are plotted by normalizing them with the peak intensity of representative lines of each spectrum; i.e., Kα lines for Fig. 7.7a–d and Heα line for Fig. 7.8e. Calibration of the spectral axis was done using spectral data from Type D target. The peak of Cl Kα line in this spectrum was identified as 2.6224 keV corresponding to Cl Kα_1 (2.62239 keV). The Kα lines appear for all types of targets, however, the amount of blue shift of Kα lines due to partial ionization of M-shell electrons is so small that all Kα lines from Cl^{+}–Cl^{8+} ions overlap and blend with each other, forming the unresolved Kα line group. Therefore, in order to identify Kα lines from higher ionization stages, this Kα line group is called "*cold* Kα *lines*" hereafter. Besides, the energy shifted Kα lines from Cl^{9+} ion is seen at 2.64 keV and that from Cl^{10+} ion is at 2.66 keV in particular for type A and B targets. The shifted Kα lines for type C and D targets are also seen although they appear to be very weak in the normalized plot. This is because the cold $K\alpha$ lines for type C and D targets are more intense than those for type A and B targets. Both cold Kα lines and Cl Heα lines are seen in the type E target, which suggests plasma formation with a large temperature gradient.

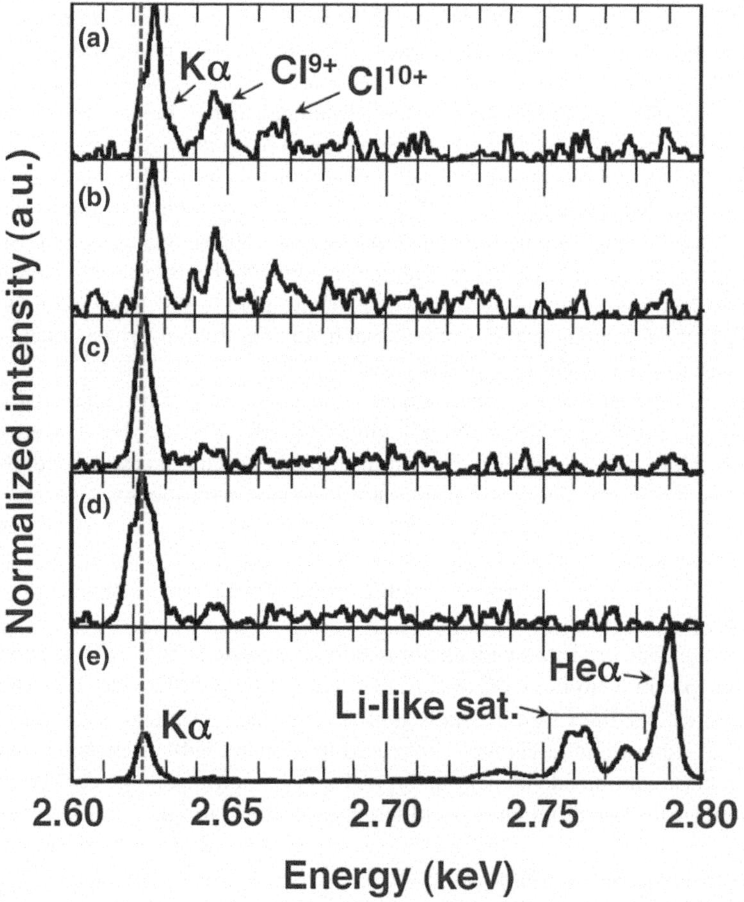

Fig. 7.7 X-ray spectra for **a** Type A (L = 50 μm). **b** Type B (L = 100 μm). **c** Type C (L = 300 μm. **d** Type D (L = 1000 μm), and **e** Type E (w/o parylene overcoat in the side of laser incidence) [43]

Figure 7.8a shows the cold Kα lines obtained with the conical crystal spectrometer for Type A-D targets. The lines are plotted in proportion to the absolute values but in arbitrary units. For reference, the energies of Cl Kα_1(2.6224 keV) and Kα_2 (2.6208 keV) are shown with arrows. Blue shift of the peaks from 2.622 to 2.626 keV is obvious with the decrease in the target mass. The same trend of the blue shift was also observed in data recorded with the toroidally curved crystal spectrometer. The cold Kα lines for plausible electron temperatures were calculated with FLYCHK code [45] as shown in Fig. 7.8b. Since Kα line from partially ionized atoms consists of several lines, thus the broadened profile is provided by their overlapping and blending in addition to the effects of Stark, thermal, and instrumental broadening. In the modeling code the instrumental broadening is a dominant factor and is set to be 1 eV, which is close to the experimental resolution of 1.3 eV at 2.6 keV. The

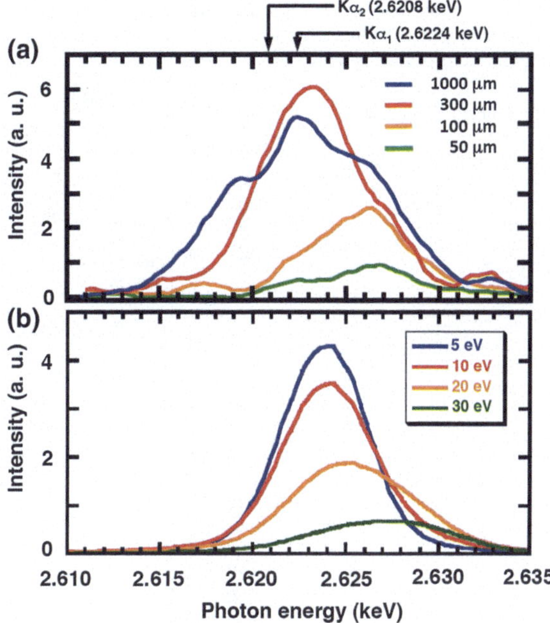

Fig. 7.8 **a** Zooming up of the cold Kα lines. *Blue shift* of the peak is obvious for the mass-reduced target. **b** Cold Kα lines calculated with FLYCHK [45] for the electron temperatures of 5–30 eV. The electron density of 1×10^{23} cm^{-3} is assumed. The *blue shift* and reduction of intensity are well replicated in comparison with the experiment [43]

electron density was chosen to be 1×10^{23}cm^{-3} in this calculation. We verified that the influence of self-absorption on the line profile is negligibly small. Although the energies of spectral peaks of FLYCHK are slightly different from those of the experiment, a similar trend of energy shift and reduction of intensity with increase in the electron temperature are well replicated. Therefore the blue shift can be attributed to the progress of ionization of Cl in an electron temperature range from 10 to 30 eV.

In this section we study the underlying physics in the ultra-fast heating process and verify the experimental results with help of the 1D and 2D particle-in-cell (PIC) code, PICLS, which incorporates relativistic binary collisions [46] and dynamic ionization.

7.7.1 1D Simulation

The laser interacts with the low density pre-plasma and produces hot electrons with energy ~194 keV, which is consistent with the ponderomotive scaling ($mc^2(1 - a^2/2)^{0.5} = 190$ keV) [35]. These hot electrons traveling further inside the target deposit energy and ionize atoms. In Fig. 7.9 the temporal evolution of the bulk electron temperature in the triple-layered target is shown. Figure 7.9a shows that in the beginning, bulk temperature rises almost uniformly inside the target ~ a few eV. The secondary heating source is the heat diffusion from the front of the target, which results in a gradient in temperature profile. With the current temperature conditions and the time scale of the heating, the drag heating is less efficient, hence the resistive

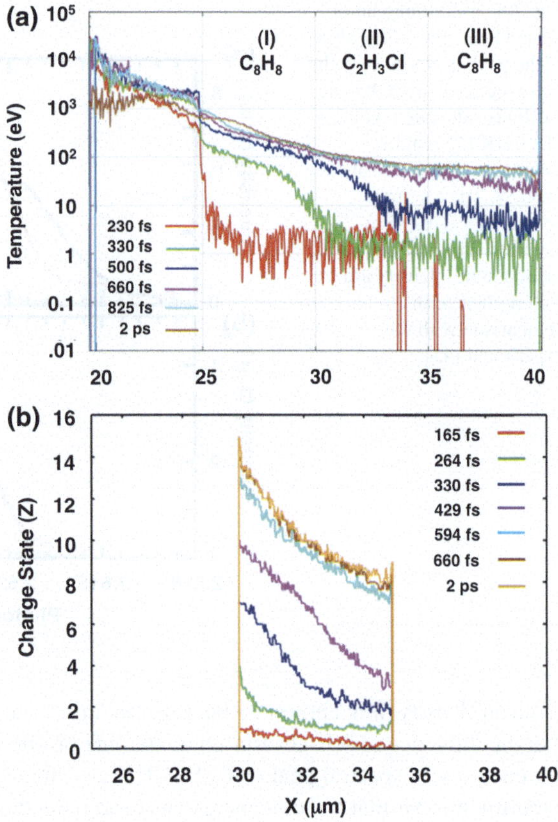

Fig. 7.9 **a** Temporal evolution of the bulk electron temperature profiles inside the triple-layered target. Temperature in the tracer layer is in the range of 60–100 eV. Laser intensity is 10^{18} W/cm^2. **b** Time evolution of chlorine ionization level in the tracer layer [43]

heating and diffusive heating are the dominant heating processes. Inside the target we see the resistive field ~6×10^9 V/m, which creates the potential of ~6 kV in a micron scale during the laser irradiation. This means the target can be heated up to keV temperatures by resistive heating if we keep irradiating the target with the laser light. However, the resistive heating will become less efficient when the hot electrons start to recirculate and supply the return current by themselves and therefore the diffusive heating appears more efficient in the current setup. The temperature jumps when a heat (ionization) wave reaches the layer from the irradiation side as seen in Fig. 7.9a. This heat wave propagates with 7 % of the speed of light.

Figure 7.9b shows the temporal evolution of chlorine ionization levels obtained by the Saha model. Here we see that the chlorine atoms continue to ionize until the laser pulse is switched off and stay unchanged after that, which is consistent with the temperature evolution. The bulk temperature stays unchanged or changed but very slowly in one-dimensional simulations, since the energy cannot diffuse to the lateral direction. Note here that Bremsstrahlung radiation loss is not significant for the current (T_e, n_e) conditions, which was confirmed by the simulation including

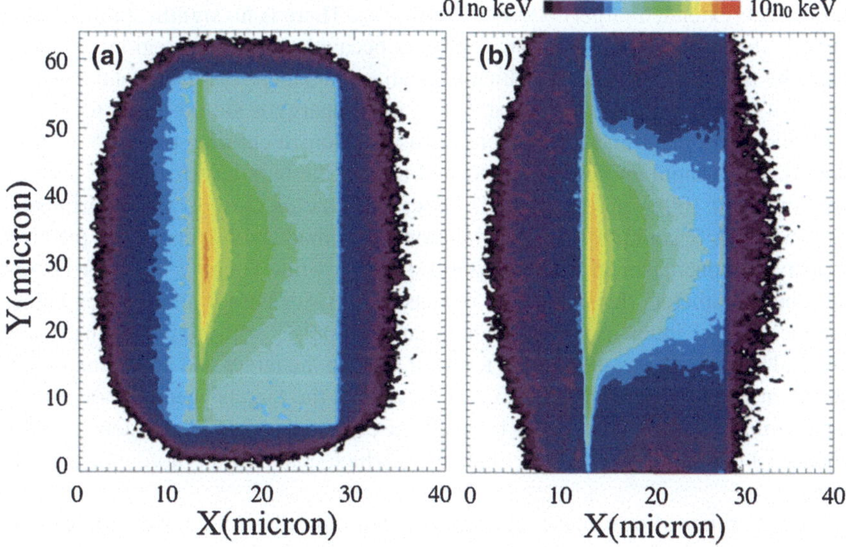

Fig. 7.10 Electron energy density profiles (in the units of keV/cm^3) at 1 ps for reduced mass (**a**) and bigger target (**b**). Here n_0 is the resolution density $n_0 = 131 n_c$ where n_c is the critical density $\sim 10^{21}$/cm^3 [43]

a simple estimation for the Bremsstrahlung radiation loss. The chlorine atoms are ionized up to higher levels near the interface of layers I-II, which is reflecting of the temperature gradient as seen in Fig. 7.9a. Note that once laser pulse is switched off at 500 fs, the bulk energy starts to thermalize inside the target, and we do not observe any further increase in bulk temperature and ionization levels of chlorine.

7.7.2 2D Simulation

Two-dimensional collisional particle-in-cell simulations show two distinct heating mechanisms occurring concurrently: uniform heating by refluxing electrons and local heating by diffusive electrons in the central region. Figure 7.10a, b shows the snapshot of electron energy density in reduced mass and bigger targets at 1 ps. In the reduced target hot electrons recirculating inside the target are likely to be confined longitudinally and laterally by electrostatic sheath fields generated at target surfaces. While the central region has the similar energy density distribution in each target, we see a slightly higher value at the irradiation surface in the layer I of the reduced mass target. This is because recirculating electrons are able to re-interact with the laser pulse during the irradiation in the smaller mass target. In Fig. 7.10a the transverse profiles of bulk temperature in the layer II for the reduced mass and extended targets at 1 ps are plot. The peak average temperature in the central region for the reduced mass target (extended target) is ~103 eV (97 eV). There is no significant difference between the two peak temperatures. This is because the central region is mostly heated by the diffusive process, which is localized and therefore does not depend on the target volume. It was found that 7 % of input energy is stored in the focal area inside the tracer layer and 0.2 % energy is used to ionize chlorine to Z ~ 5–6. Also about 0.3 % of input energy was deposited to heat the bulk electron to the average temperature of 100 eV in the layer II. These numbers are close to the experimental observation. A few additional 2D simulations were made with increases in the laser intensity and decreases in the laser pulse duration while keeping the laser energy and other parameters the same. The increased intensities are 4 and 8 times higher than the original intensity. It was found that laser pulses with higher intensity and shorter duration are less efficient in terms of heating the target and achieving higher ionization degree of chlorine. Also it is noted that the targets' geometrical effects become more significant in higher intensity cases, namely, the higher intensity pulse produces a larger number of hotter electrons, which could reflux more uniformly in the target.

For both types of targets the chlorine ionization profiles follow the similar trend as the average temperature and are similar if one looks at the region corresponding to the focal area of the tracer layer, with a peak at $Z \sim 11$. Nevertheless, the difference in average Z profiles near the boundaries can be attributed to the target volumetric effect. Figure 7.11 shows the distribution of chlorine ion stage abundances in the layer II. A twin-peak distribution for both targets is seen. This suggests the presence of two characteristic temperatures, which are associated with different regions. Actually the

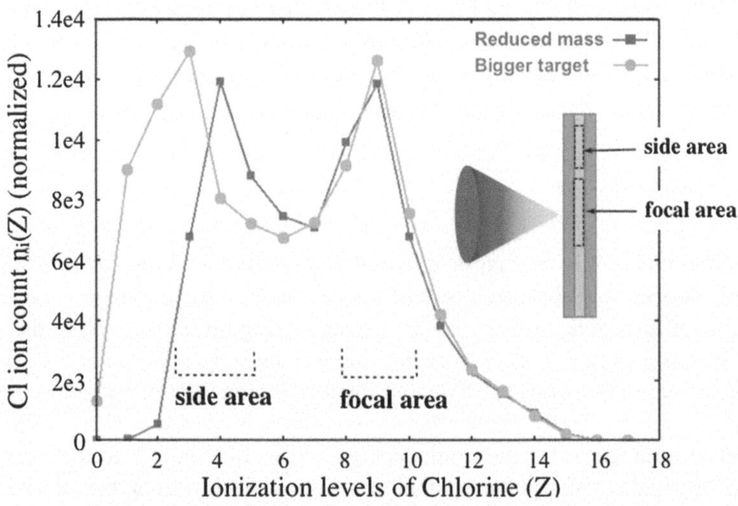

Fig. 7.11 Distribution of chlorine ion number in the respective ionization levels in the tracer layer for reduced-mass and bigger-mass targets. A twin-peak distribution for both targets is seen, suggesting the presence of two characteristic temperatures associated with two separate regions heated by different mechanisms [43]

lower Z ($Z < 6$) ions are mostly observed outside of the focal regions, i.e. side areas of target indicated in Fig. 7.11, while the higher Z ($Z > 8$) ions are found in the central region. FLYCHK calculations show that population of Cl^{8+} and Cl^{9+} ions, as opposed to the source ions for shifted Kα lines from Cl^{9+} and Cl^{10+}, become highest if the electron temperature is ranging from 85 to 100 eV for the electron density of $2 \times 10^{23}\,\text{cm}^{-3}$. This temperature agrees quite well with the PIC simulations. Namely the shifted Kα lines are likely emitted from the hot region ($T_e \sim 90\,\text{eV}$) in the tracer layer whose temperature profiles are independent on the target mass.

This twin peak distribution suggests the presence of two different heating mechanisms that are going on simultaneously. The first one is a uniform heating by refluxing hot electrons, which are recirculating inside the target. This heating is more effective for the smaller target since there we see the peak appearing at a higher ionization level ($Z = 4$) than for the extended target ($Z = 3$). The second mechanism is the diffusive heating at the central area of the front surface. The diffusive heating is a local process and depends on the local temperature gradient. Since the diffusive heating does not depend on the target volume, there is no significant difference in its importance between smaller and bigger targets. We see the higher peak at $Z = 9$ due to a higher temperature achieved by the diffusive heating. Note here that the observed blue-shift of the peak of cold Kα lines for smaller targets, see Fig. 7.8a, is consistent with the up-shift of the ionization peak at lower Z(3–4) accompanying a change of the target size.

7.8 Conclusions

In this paper we have described the methods, which can be used to access Warm Dense Matter and High Energy Density states, and how it is possible to diagnose such states. Such a study is interesting for many branches of science ranging from material science, to planetology, astrophysics, and inertial confinement fusion. The beginning of operation of new laser facilities (like LMJ/PETAL in France) will allow creating even larger samples, thereby easing experimental measurements and allowing obtaining reduced error bars. Progress in theoretical description and numerical simulations are also needed to describe such complicated states of matter.

Acknowledgments This work has been carried out with the support of the cost Action MP1208 and of the French cluster LAPHIA (contract ANR-10-1DEX-03-02).

References

1. R. Fabbro, E. Fabre et al., Phys. Rev. A, **26**, 2289 (1982); R. Fabbro, C.E. Max, E. Fabre, Phys. Fluids, **28**, 2580 (1985); R. Fabbro, et al., Phys. Fluids **28**, 3414 (1985)
2. T. Nakajima et al., Nature **378**, 463 (1995); B. Oppenheimer et al., Science **270**, 1478 (1995)

3. Ya. Zeldovich, Yu. Raizer, Physics of Shock Waves and High Temperature Hydrodynamic Phenomena (Academic Press, New York, 1967)
4. F. Cottet et al., Appl. Phys. Lett. **47**, 678 (1985)
5. M. Koenig, D. Batani et al., Phys. Rev. E **50**, R3314 (1994); M. Koenig, D. Batani et al., Phys. Rev. Lett. **74**, 2260 (1995); M. Koenig, et al., Las. Part. Beams **10**, 573 (1992)
6. Y. Kato et al., Phys. Rev. Lett. **53**, 1057 (1984)
7. R.M. Stevenson, C. Danson, I. Ross et al., Opt. Lett. **19**, 363 (1994)
8. L.B. Da Silva, P. Celliers, G.W. Collins, K.S. Budil, N.C. Holmes, T.W. Barbee Jr., B.A. Hammel, J.D. Kilkenny, R.J. Wallace, M. Ross, R. Cauble, A. Ng, G. Chiu, Absolute equation of state measurements on shocked liquid deuterium up to 200 GPa (2 Mbar). Pys. Rev. Lett. **78**, 483 (1997)
9. S. Fahy, S.G. Louie, Phys. Rev. B **36**, 3373 (1987)
10. A.L. Ruoff, H. Luo, J. Appl. Phys. **70**, 2066 (1991)
11. H.K. Mao, P.M. Bell, Science, **200**, 1145 (1978); H.K. Mao, R.J. Hemley. Nature **351**, 721 (1991)
12. F.P. Bundy, Physica A, **156**, 169 (1989); F.P. Bundy, J. Chem. Phys. **38**, 631 (1963); F.P. Bundy, H.M. Strong, R.H. Wentorf. J. Chem. Phys. **10**, 213 (1973)
13. M. Grumbach, R. Martin, Phys. Rev. B **54**, 15730 (1996)
14. T. Sekine, Appl. Phys. Lett. **74**, 350 (1999)
15. L. Benedetti et al., Science **286**, 100 (1999)
16. S. Scandolo et al., Phys. Rev. B **53**, 5051 (1996)
17. A. Cavalleri et al., Europhys. Lett. **57**, 281 (2002)
18. In this regime, the most complete EOS are the SESAME tables developed at the Los Alamos Laboratory (SESAME Report on the Los Alamos Equation-of-State library, Report No. LALP-83-4, T4 Group LANL, Los Alamos, 1983)
19. M. Ross, Rep. Prog. Phys. **48**, 1 (1985)
20. S. Eliezer, A. Ghatak, H. Hora, Equation of State: Theory and Applications, (Cambridge Univ. Press, Cambridge, 1986)
21. T. Guillot, Science **286**, 72 (1999)
22. S. Saumon, G. Chabrier et al., Astrophys. J. Suppl. **99**, 713 (1995)
23. J.E. Connerney et al., J. Geophys. Rev. **92**, 15329 (1987)
24. N.F. Ness et al., Science, **240**, 1473 (1989); N.F. Ness, et al., Science **233**, 85 (1986)
25. D. Batani, H. Stabile, M. Tomasini, G. Lucchini, A. Ravasio, M. Koenig, A. Benuzzi-Mounaix, H. Nishimura, Y. Ochi, J. Ullschmied, J. Skala, B. Kralikova, M. Pfeifer, Ch. Kadlec, T. Mocek, A. Präg, T. Hall, P. Milani, E. Barborini, P. Piseri, Hugoniot data for carbon at megabar pressures. Phys. Rev. Lett. **92**, 065503 (2004)
26. M.N. Pavlovskii, V.P. Drakin, JETP Lett. **4**, 116 (1966)
27. M.N. Pavlovskii, Sov. Phys. Solid State **13**, 741 (1971)
28. W.J. Nellis et al., J. Appl. Phys. **90**, 696 (2001)
29. W.H. Gust, Phys. Rev. B **22**, 4744 (1980)
30. S.P. Marsch (ed.), *LASL Shock Hugoniot Data* (University of California, Berkeley, 1980), pp. 28–51
31. S. Baton, D. Batani et al., Relativistic electron transport and confinement within charge-insulated, mass-limited targets. High Energy Density Phys. **3**, 358 (2007)
32. Y. Fukuda, A. Faenov et al., Generation of X rays and energetic ions from superintense laser irradiation of micron-sized Ar clusters. Laser Part. Beams **22**, 215 (2004)
33. P. Hakel, K. Eidmann, F. Pisani et al., X-ray line emissions from tamped thin aluminum targets driven by subpicosecond-duration laser pulses. High Energy Density Phys. **5**, 35 (2009)
34. P.K. Patel, R. Stephens et al., Isochoric heating of solid-density matter with an ultrafast proton beam. Phys. Rev. Lett. **91**, 125004 (2003)
35. S.C. Wilks, W.L. Kruer, M. Tabak, A.B. Langdon, Phys. Rev. Lett. **69**, 1383–1386 (1993)
36. S. Atzeni, M. Tabak, Plasma Phys. Controlled Fusion **47**, B769–B776 (2005)
37. H. Alfven, Phys. Rev. **55**, 425–429 (1939)
38. A.J. Kemp, Y. Sentoku, V. Sotnikov, S.C. Wilks, Phys. Rev. Lett. **97**, 235001-1-4 (2006)

39. S. Fujioka, H. Takabe, N. Yamamoto, D. Salzmann, F. Wang, H. Nishimura, Y. Li, Q. Dong, S. Wang, Y. Zhang, Y.J. Rhee, Y.W. Lee, J.M. Han, M. Tanabe, T. Fujiwara, T. Nakabayashi, G. Zhao, J. Zhang, K. Mima, Nat. Phys. **5**, 812–815 (2009)
40. W. Schwanda, K. Eidmann, Phys. Rev. Lett. **69**, 3507–3510 (1992)
41. J. Colgan, J. Abdallah, A.Ya. Faenov, S.A. Pikuz, E. Wagenaars, N. Booth, O. Culfa, R.J. Dance, R.G. Evans, R.J. Gray, T. Kaempfer, K.L. Lancaster, P. McKenna, A.L. Rossall, I. Yu. Skobelev, K.S. Schulze, I. Uschmann, A.G. Zhidkov, N.C. Woolsey, Phys. Rev. Lett. **100**, 125001-1-5 (2013) and references therein
42. H.S. Park, D.M. Chambers, H.-K. Chung, R.J. Clarke, R. Eagleton, E. Giraldez, T. Goldsack, R. Heathcote, N. Izumi, M.H. Key, J.A. King, J.A. Koch, O.L. Landen, A. Nikroo, P.K. Patel, D.F. Price, B.A. Remington, H.F. Robey, R.A. Snavely, D.A. Steinman, R.B. Stephens, C. Stoeckl, M. Storm, M. Tabak, W. Theobald, R.P.J. Town, J.E. Wickersham, B.B. Zhang, Phys. Plasmas **13**, 056309-1-10 (2006)
43. H. Nishimura, R. Mishra, S. Ohshima, H. Nakamura, M. Tanabe, T. Fujiwara, N. Yamamoto, S. Fujioka, D. Batani, M. Veltcheva, T. Jafer, T. Kawamura, T. Sentoku, R. Mancini, P. Hakel, F. Koike, K. Mima, Phys. Plasmas **18**, 022702-1-9 (2011)
44. P.K. Patel, A.J. Mackinnon, M.H. Key, T.E. Cowan, M.E. Foord, M. Allen, D.F. Price, H. Ruhl, P.T. Springer, R. Stephens, Phys. Rev. Lett. **104**, 085001-1-4 (2010)
45. H.K. Chung, W.L. Morgan, R.W. Lee, J. Quant. Spectrosc. Radiat. Transf. **81**, 107 (2003)
46. Y. Sentoku, A.J. Kemp, J. Comp. Phys. **227**, 6846 (2008)
47. H.M. Milchberg et al., Phy. Rev. Lett. **61**(20), 2364 (1988)
48. K. Eidmann, J. Meyer-ter-Vehn, T. Schlegel, S. Huller, Phys. Rev. E **62**, 1202 (2000)
49. L. Spitzer, *The Physics of Fully Ionised Gases* (Wiley Interscience, New York, 1962)

Chapter 8
Laser-Plasma Particle Sources for Biology and Medicine

Antonio Giulietti, Giancarlo Bussolino, Lorenzo Fulgentini, Petra Koester, Luca Labate and Leonida A. Gizzi

Abstract Ultrashort, intense laser pulses can drive in plasmas small sized linear accelerators (Laser-Linac's) of high energy elementary particles. These novel devices are facing a continuous, fast progress making them suitable alternatives to conventional linacs in many applications. Among them, cancer therapy may have by far the highest social impact at a global level. This paper is aimed at giving an updated overview of the scientific and technological effort devoted worldwide to the optimization of the laser acceleration technology in order to fulfill the clinical requirements. Here we discuss both ion and electron acceleration considering the different, challenging problems to be solved in each case. Current studies on radiobiology already in progress in many labs with the existing laser-based sources of particles are also described. The overall scenario in the field appears extremely exciting, and promises rapid, effective development.

A. Giulietti (✉) · G. Bussolino · L. Fulgentini · P. Koester · L. Labate · L.A. Gizzi
Intense Laser Irradiation Laboratory, Istituto Nazionale di Ottica,
CNR Campus, via Moruzzi, 56124 Pisa, Italy
e-mail: antonio.giulietti@ino.it

G. Bussolino
e-mail: giancarlo.bussolino@ino.it

L. Fulgentini
e-mail: lorenzo.fulgentini@ino.it

P. Koester
e-mail: petra.koester@ino.it

L. Labate · L.A. Gizzi
INFN, Sezione di Pisa, Italy
e-mail: luca.labate@ino.it

L.A. Gizzi
e-mail: leonidaantonio.gizzi@ino.it

K. Yamanouchi et al. (eds.), *Progress in Ultrafast Intense Laser Science XII*,
Springer Series in Chemical Physics 112, DOI 10.1007/978-3-319-23657-5_8

8.1 Introduction

High-field photonics [1] is one of the most exciting branches of ultrashort intense laser science. Powerful femtosecond lasers, coupled with suitable focusing optics, can shoot into matter optical "bullets" whose transverse size (the spot size) is comparable with the longitudinal size (the pulse length) and whose photon density is of the order of 10^{27} phot/cm^3. The oscillating electric field in the bullet volume can exceed 10^{12} V/cm, much higher than atomic fields. When interacting with matter, such pulses are able to ionize atoms in a time of the order of a single optical cycle, i.e. in a fs or less. Free electrons are then moved by the oscillating electric field to relativistic quiver velocities, so opening the novel field of investigation commonly referred to as *optics in the relativistic regime* [2].

In this framework, we can roughly distinguish between two levels of laser intensity. At a lower level, the electric field can be just comparable with the one requested for the ionization of the medium. This condition produces a strong modification of the laser pulse parameters, including the oscillation frequency of the e.m. field. This case has been studied by several authors. In particular, Le Blanc et al. [3] reported one of the first clear observations of spectral blue shifting of a femtosecond laser pulse propagating in a dense gas. Interestingly, a well defined spectral shift was also measured after anomalous propagation in an overdense plasma slab [4]. A number of papers report on spectral effects in conditions of interest for laser driven electron acceleration. Koga et al. observed blue shift up to 40–50 nm, which was attributed to a combined effect of ionization and filamentation [5]. Giulietti et al. observed 25 nm peak blue shift with modulated spectral tails extending up to 100 nm, mostly attributed to self-phase modulation [6]. Recently [7], observation of extreme blue shifting has been reported for a 65-fs, 800-nm, 2-TW laser pulse propagating through a nitrogen gas jet, experimentally studied by 90° Thomson scattering. Time-integrated spectra of scattered light show unprecedented broadening towards the blue which exceeds 300 nm. Images of the scattering region provide for the first time a space- and time-resolved description of the process leading quite regularly to such a large upshift. The mean shifting rate was as high as $d\lambda/dt \approx 3$ Å/fs, never observed before. In principle this effect allows the spectrum of ultrashort laser pulses to be shifted and tuned, with possible applications in many fields. It has to been noticed that relevant laser frequency shifts due to ultrafast ionization are in general observed at intensity below the one suitable for laser acceleration. In particular the extreme shift described in [7] was observed in an acceleration experiment but during the late propagation of the laser pulse in a gas jet, when its intensity had been reduced by energy depletion and beam defocusing.

At higher laser intensities, ionization time is much shorter than the usual pulse duration (a few tens of femtoseconds) and almost all the pulse propagates in a fully ionized medium. In this condition the e.m. field is able to excite, via ponderomotive displacement in a plasma of suitable density, electron waves of very large amplitude, whose longitudinal electric field has the typical properties of an accelerator cavity, able to accelerate free electrons up to relativistic kinetic energies. The suitable plasma density depends on the regime of acceleration chosen but usually the plasma working density is well below the critical density $n_c = (m_e/4\pi e^2)\omega_L^2$, where e and m_e are electron charge and mass respectively, ω_L is the laser light pulsation. Local acceleration fields in these electron plasma waves are at least 3 orders of magnitude higher than in the cavities of Radio-Frequency driven ordinary Linacs. This means for example that multi-MeV electrons can be produced in a millimetric laser path rather than in meter-long RF cavity. A striking feature of the laser-produced electron bunches is their ultrashort duration, typically of the order of 1 ps, more than 10^6 times shorter than bunches produced with RF technology. Possible relevant consequences of this feature will be discussed below.

Since the original idea of Laser Wake-Field Acceleration [8] and the advent of the decisive CPA laser technology [9], a number of schemes for laser driven acceleration of electrons in plasmas have been proposed and studied, some of which were successfully tested. New experimental records have been reported in the recent literature, in terms of the maximum electron energy achieved, the minimum energy spread, as well as maximum collimation, stability, and so on. These records are in general obtained with lasers of outstanding performances and/or with very sophisticated methods hardly applicable for practical uses. On the other hand, many labs are intensively working on scientific and technological innovations aimed at demonstrating that reliable laser-based devices can be built which are able to produce electron beams fulfilling requirements of specific applications. A major task is addressed to the possible clinical use of electron Laser linacs and their potential advantages with respect to the existing RF-linacs operating today for millions of daily hospital treatments in the world.

Laser driven electron acceleration through excitation of plasma waves acts on free electrons already available in a plasma. In general the primary interaction of the laser field is with electrons (either bound or free), while action on massive particles (protons and ions) needs the intermediate role of electrons. For this reason, though evidence of the effect of the laser field on the ion velocity was found as early as high power lasers entered the laboratory, the first relevant effects on ion acceleration were observed, in the fusion research context, with powerful CO_2 lasers [10, 11]. These latter in fact, due to their large wavelength (10-μm) can induce huge electron quiver velocities on plasma electrons.

Historically, ion acceleration in plasmas was proposed before the invention of optical lasers, as early as 1956 [12] and initially tested with electron beams propagating in plasmas. Apart from initial observation related to fusion studies with infrared CO_2 lasers cited above, the laser driven ion acceleration studies with optical lasers could really start only after some decisive breakthrough towards high peak power lasers, like mode-locking (ML) for picosecond pulses and chirped pulse amplification (CPA) for femtosecond pulses [9]. About 1-MeV ions were produced in the early Nineties with picosecond laser pulses [13]. Since then, an impressive progress towards higher kinetic energies was continuously driven by both innovation in laser technology and better comprehension of the complex physics involved in the ion acceleration processes. Several proposals raised for a variety of schemes of laser-matter interaction at ultra-high (ultra-relativistic) intensities able to drive protons and

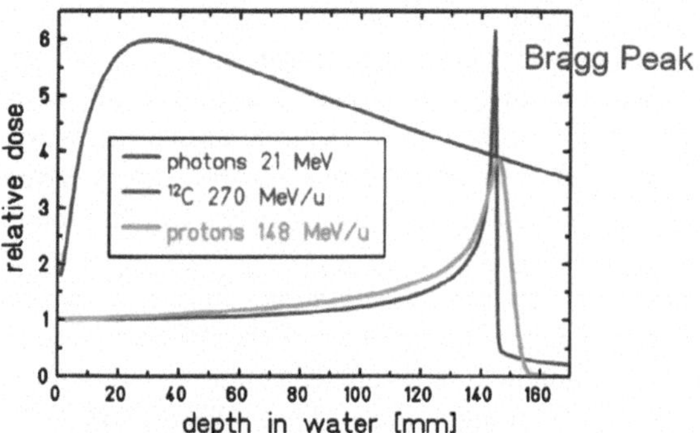

Fig. 8.1 Relative dose deposition versus depth in water for three kinds of ionizing agents, each one with a specific energy

light ions to near-relativistic energies. Most of them can be attributed either to *target normal sheath acceleration* (TNSA) or *radiation pressure dominated acceleration* (RPDA), shortly described in Sect. 2.2.

Also in the case of protons/ions accelerated with laser-based techniques, clinical applications (chiefly hadron-therapy of deep tumors) are a major objective, considering the peculiar character of energy deposition of hadrons in a medium. Figure 8.1 clearly shows that, treating a tumor at 15 cm depth, monoenergetic protons and Carbon ions of suitable kinetic energy deliver most of the dose in a thin layer (Bragg peak) around the tumor site, while monochromatic gamma-rays leave a lot of energy inside healthy tissues before and after the tumor with possible damages on these latter tissues. This drawback has been strongly reduced with modern configurations allowing multi-beam irradiations at different angles. It has to be considered that gamma-ray (or hard X-ray) treatments are still more than 95 % of the current treatments worldwide, while hadron therapy is limited due to the size and cost of proton/ion accelerator plants. That is why laser-based techniques of acceleration have been indicated quite early as a possible way to access hadron treatment of cancer with reduced cost and plant size [14].

In general, considering the present state of the art, we can say that laser-driven acceleration to kinetic energies suitable for radiotherapy of cancer is well consolidated in the case of electrons and bremsstrahlung photons (with bunches delivering the requested dose). Effort is being invested towards achievement of corresponding energies for protons and light ions. Time for technological and commercial alternative with existing Hospital electron-Linac's as well as with rare, huge plants already operating hadron therapy, may not be so far. In the case of electrons most of the work to be done, in order to achieve clinical standards, has to address the control of the electron energy, as well as stability and reliability of the laser-linac. In the case of protons and light ions the work to be done still includes the identification of an acceleration regime able to produce particles of suitable energy and energy spread in bunches delivering the right dose.

However, there is a major scientific issue which has to be addressed from now, concerning potential radiobiological effects of the extremely different duration of bunches produced by laser with respect to bunches produced by conventional accelerators. A factor exceeding 1,000,000 is involved, from μs to sub-ps timescale. The ultrashort duration of laser-produced particle bunches may involve unexpected consequences for cancer therapy. In fact, it is not known if delivering the same dose with particles of the same kinetic energy but at much higher instantaneous dose-rate may lead to a different tissutal effects with possible consequences on therapeutic strategy and protocols [15]. From the physical point of view we can expect that the extreme particle density we can produce in a bunch with laser acceleration could produce some "collective effect" which cannot be described by the usual single-particle Monte Carlo simulation. In other words it is possible that each ultradense bunch of electrons could produce not only the statistic sum of the effects of each low-LET particle but also some high-LET effect due to the total charge involved. If this would be true, the biological action could not only concern DNA but also some structural cellular feature, like membrane. This major issue, in turn, calls for a dedicated research on radiobiological effects to be performed with the ultrashort particle bunches produced by laser technology. It is evident that such a research also has a high conceptual value since it enables, for the first time, the investigation of very early processes occurring in the timescales of physical, chemical, biological responses of the living matter to ionizing radiation [16].

In this work we will give a concise description of the state of the art and perspectives of the laser technologies currently being developed to address implementation of compact particle accelerators for biological research and clinical uses. Section 8.2 will be devoted to laser-based hadron acceleration and their use in radiobiology with an evaluation of their progress in the perspective towards radiotherapy of tumors. Need of upgrading existing lasers will be also considered. Section 8.3 will shortly describe different kinds of presently available laser Linac's with particular attention to the clinical requirements, efficiency and reliability of the devices. Interest to sub-relativistic electron sources will also be discussed. In Sect. 8.4 we will consider the main features of radiobiology with laser-plasma electron source, including dosimetry, in vivo tests and numerical simulations. Possibility of investigating very early effects arising from ultrashort ionizing pulses at nanometric scale will be discussed.

8.2 Radiobiology with Protons and Ions from Laser-Based Sources

8.2.1 Present RF-Driven Versus Future Laser-Driven Devices (Protons and Ions)

Considering the great advantages of hadron therapy and the fast progress in laser-based techniques, about 10 years ago, a few interesting and detailed proposals were published for using laser-driven hadron accelerators in medicine for radiological treatment of cancer [14, 17]. These proposals were strongly motivated by the huge

Fig. 8.2 Overview of the CNAO synchrotron, Pavia, Italy

size and cost of present conventional ion accelerators, mostly cyclotron and synchrotron devices, few tens of which are operating nowadays worldwide in a clinical background. As an example, Fig. 8.2 shows the overview of the Centro Nazionale di Adroterapia Oncologica (CNAO) synchrotron operating for clinical use in Pavia (Italy) [18]. Interesting information on principles of proton-therapy and treatment options in the advanced Northeast Proton Therapy Center (Boston, USA), can be found in [19].

After the first proposal for using hadron particles for radiotherapy of cancer [20] and several pioneering experimental tests [21], hadron therapy was occasionally performed inside accelerator facilities devoted to high energy physics, until the opening (1990) of a first clinical center equipped with a proton accelerator facility at Loma Linda Hospital in California (USA). Since then, the number of similar centers grew regularly year by year worldwide with a huge capital investment. The most important facilities operating in the year 2011 are listed in Table 8.1, in the order of their opening year.

Though RF-based devices have faced an impressive progress, mostly in the synchrotron configuration [22], typical acceleration gradients remain of the order of 1MeV/m, so that the typical diameter of an accelerator ring is several tens of meters for energies of clinical interest, namely $\mathrm{E} \approx 100$–$400\ \mathrm{MeV/u}$, with severe costs involved [23]. Additional high costs and large spaces are requested by the very heavy gantry systems necessary to guide the particle beam onto the patient body from the right direction(s) and focus it with a millimeter precision [24].

With such a strong motivation, research on laser-based proton acceleration has been considerably supported in the last decade, mostly in the direction of achieving the challenging performances requested by the clinical standards. According to Fig. 8.1 and Table 8.1, we can see that a usable device for cancer therapy needs to produce 200–250 MeV protons and/or 400–450 MeV/u carbon ions. In order to really profit of the Bragg peak, no more than 1 % energy bandwidth is requested. Further, to release a dose of therapeutic interest in a reasonable time, more than 10^{10}

Table 8.1 The most important clinical centers for hadron therapy operating in 2011

1990	Loma Linda	USA	Protons	250 MeV
1994	Himac Chiba	JPN	Carbon	800 MeV/u
1994	NCC, Kashiwa	JPN	Protons	235 MeV
2001	HIBMC, Hyogo	JPN	Protons	230 MeV
2001	PMRC, Tsukuba	JPN	Protons	250 MeV
2001	NPTC, MGH, Boston	USA	Protons	235 MeV
2003	Shizuoka	JPN	Protons	235 MeV
2004	MPRI, Bloomington	USA	Protons	200 MeV
2004	WPTC, Zibo	CHN	Protons	230 MeV
2006	MD Anderson Houston	USA	Protons	250 MeV
2006	FPTI, Jacksonville	USA	Protons	230 MeV
2007	NCC, Ilsan	KOR	Protons	230 MeV
2009	ProCure, Oklahoma City	USA	Protons	230 MeV
2009	RPTC, München	GER	Protons	250 MeV
2009	HIT, Heidelberg	GER	Protons, Carbon	430 MeV/u
2010	UPenn, Philadelphia	USA	Protons	230 MeV
2010	CNAO, Pavia	ITA	Protons, Carbon	430 MeV/u
2010	WPE, Essen	GER	Protons	230 MeV
2010	CPO, Orsay	FRA	Protons	230 MeV
2010	PTC, Marburg	GER	Protons, Carbon	430 MeV/u
2010	Gunma, Maebashi	JPN	Protons	400 MeV/u
2010	HUPBTC, Hampton	USA	Protons	230 MeV
2010	SJFH, Peking	CHN	Protons	230 MeV

Source http://www.klinikum.uni-heidelberg.de/Therapy-centers-in-the-world

particle/s have to reach the tissue under treatment. None of these performances has been achieved so far with laser techniques. Some of them seem still hard to achieve with existing lasers or even with the new generation lasers, at least in a configuration practically usable in a hospital context. In the following we will consider several options which are currently investigated or could be in the near future with upgraded laser systems.

8.2.2 Laser-Based Ion Acceleration Schemes Suitable for Medical Applications

A high power laser primarily acts with e.m. field on electrons (first bound, then free after ionization), while action on massive particles (protons and ions) needs the intermediate role of these electrons. This scenario has been described in many works after the advent of powerful laser systems and has been recently reviewed within

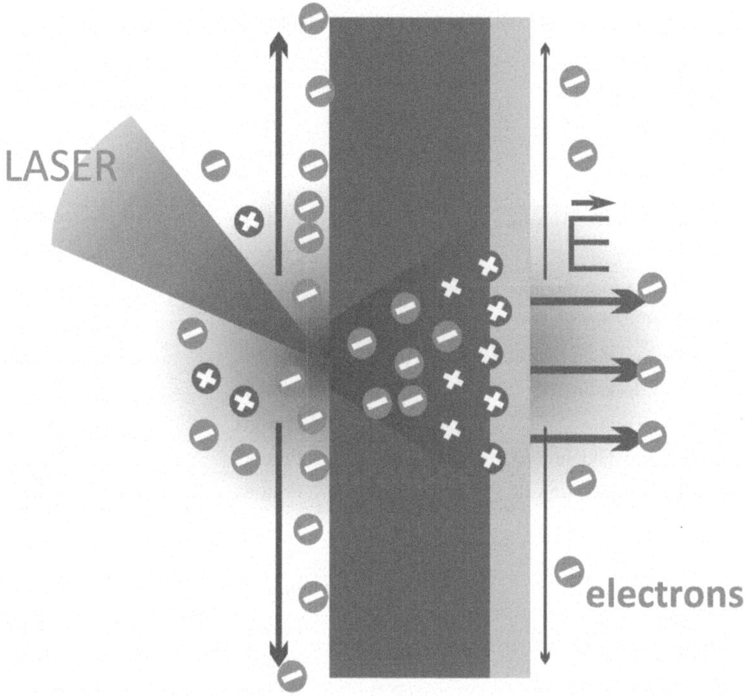

Fig. 8.3 Sketch of laser induced charge dynamics preceeding proton acceleration by TNSA

the correct theoretical background by Mulser and Bauer [25]. More recent review papers specifically devoted to laser-driven ion acceleration includes [26, 27], this latter more addressed to applications.

Let's briefly consider here the main schemes currently under study for the acceleration of protons and ions via laser-matter interaction in view of their possible application to the medical purposes, first of all radiotherapy of cancer. We will refer to Fig. 8.3 as a basic geometry involving three physical elements, namely the *laser pulse* focused on a *target* and generating a *proton beam*. Each of the three elements, together with the interaction geometry, will be characterized by a set of parameters whose combination will define the physical regime of interaction and consequently the regime of proton acceleration. Notice that Fig. 8.3 is merely representative and the actual geometry can be considerably different, e.g. available protons could escape also from the opposite side of the target.

In order to be of therapeutic interest the proton (ion) beam has to release more than 10^{10} quasi-monoenergetic particles per second, with a kinetic energy >200 MeV (400 MeV/u). This challenging performances need probably a laser pulse peak power exceeding the PW at a pulse repetition rate of the order of kHz and a high speed target-refreshing system, with a high degree of shot-to-shot reproducibility of

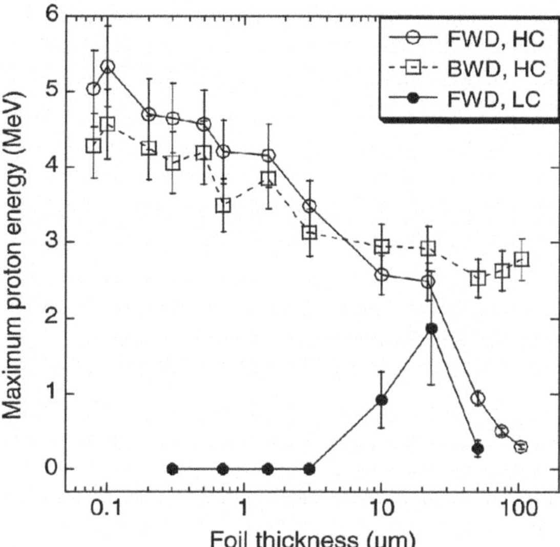

Fig. 8.4 Maximum proton energy versus target thickness for high-contrast (HC) and low-contrast (LC) laser pulses (Ceccotti et al. [31])

the particle beam parameters. Of course a crucial role will be played by the physical processes involved in the acceleration regime.

The *target normal sheath acceleration* (TNSA) regime has been the most studied so far, experimentally, theoretically and numerically. A TNSA model was firstly proposed by Wilks et al. in 2001 [28] to interpret the first experimental observations [29], of high energy protons escaping the rear side of the laser irradiated target [30] like in the sketch Fig. 8.3. This process is described as due to the extremely high electric field generated by many relativistic electrons escaping the rear surface and forming a cloud parallel to that surface.

A rich crop of experimental data basically confirmed the model and enriched it with a number of variants, including laser pulse and target optimization. Target thickness was proved to be a crucial parameter, leading to either higher proton energy with thinner target, provided the *contrast* of the main laser pulse over the *pre-pulse* (see Fig. 8.5) was very high [31], or determining an optimum thickness for a given finite contrast [32], as shown in Fig. 8.4. The data plots also show that with ultra-high contrast and thin foil targets the proton energy has about the same trend for protons coming either from the front or from the rear face [31].

As for a possible accelerator based on thin foils, high repetition rate requested by clinical uses would require sliding tape or similar devices. Another parameter to be exploited is the target size (mass), which led to a series of experiments with "mass-limited" targets. In principle, reducing the target surface should result in a denser electron cloud and higher ion accelerating field. This assumption is well confirmed by numerical simulations but only partially by experiments. High rate injection of such "pellets" would be an additional problem for an accelerator operating at high rep-rate.

Apart for some "exotic" targets or sophisticated configurations [26, 27], proton spectra produced by TNSA show a broad, thermal-like energy spectrum. This feature risks to vanish the advantages of the Bragg peak in deep energy deposition, which is the strongest motivation for hadron therapy. Most of the energy broadening in the TNSA acceleration is due to the initial distribution of protons in a wide region where the accelerating field varies considerably. An effort at designing special targets (e.g. with small dots of proton-rich material on surface or "grating targets" [33]) is currently in progress with a partial but encouraging success. At the same time several kind of passive filters able to reduce the outcoming proton spectrum are tested. It has to be considered, however, that any kind of passive particle filtering will introduce an additive radioactivation trouble in a clinical context.

Given the current limitations of maximum energy and energy spread of TNSA based ion sources, a possible approach consists in the use of hybrid structures including a TNSA based source, acting as an injector, and a post-acceleration assembly to select and boost the energy to the levels needed for clinical tests. In fact, in the TNSA regime, the proton beam is characterized by an exponential energy spectrum, with a cutoff energy and broad angular divergence. Conventional transport lines are being designed [34] to perform energy selection and beam collimation. These transport and selection lines are based on compact solenoid and magnetic quadrupoles. To increase the maximum available energy, injection of protons into a small linac has been proposed for post-acceleration. Realistic start-to-end simulations of these schemes, based on actual TNSA performances, show that this approach may lead to a prompt and profitable exploitation of currently available laser-driven proton and ion sources.

Let's now consider *radiation pressure dominated acceleration* (RPDA). Pressure of e.m. radiation on an interface of two media is a classical, non-trivial topic [35] still under investigation [36], including possible applications, ranging from pushing aircrafts to accelerating particles with powerful laser pulses. This latter option has been widely discussed and partially tested experimentally for proton acceleration in the last decade [26, 27]. Without going in detail, we will briefly consider its potential for medical applications, as well as some of its drawbacks. In order to avoid the misleading crop of terms used in the literature, like "hole boring", "laser piston", "sweeping effect", "light sailing", etc., let us simply distinguish between the two limit cases of thick and thin target, respectively.

In this context, we can consider "thick" a target much thicker than the one for which ion acceleration can occur by the direct space-charge effect. In this condition interpretation of experimental results is quite tricky since the action of the radiation pressure is mixed and somehow confused with the action of the shock wave propagating in the target material. The shock wave in turn can be partially pushed by light pressure at very high irradiance. This scenario is further complicated by the presence of a plasma in front of the target before the arrival of the main laser pulse (pre-plasma), unavoidable with thick targets. All this makes presently difficult to evaluate the potential of thick target technique for medical applications, in spite of a large crop of models and results in a variety of conditions.

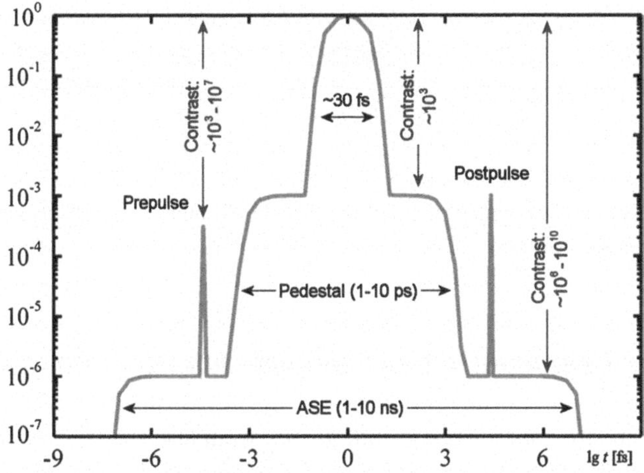

Fig. 8.5 Time evolution of parasitic laser emission before and after the mail pulse

The situation is different with "thin" targets, i.e. so thin that the laser pulse *passes through* the target and ions are accelerated *during* the pulse itself. This regime is extremely promising for applications which need ions of very high energy with control of their spectrum. However, for a clear and complete test of the various scaling laws published in the last years, laser intensity $>10^{23}$ W/cm^2 is requested. This condition demands then a general upgrading of the existing lasers as discussed in the next paragraph. There are plenty of other possible acceleration mechanisms of ions from ultra-high intensity laser interaction with matter which have been proposed and (some of them) very preliminary tested [26, 27]. Their discussion is out of the purpose of this work. We only observe that this variety of proposals proves that this field is extremely active and productive scientifically, but probably still not mature for practical applications, in particular for the highly demanding clinical application.

As already anticipated above, in most of the laser-based ion acceleration schemes a crucial role is played by the laser *pulse contrast*, more exactly by the ratio between the main pulse peak *power* and the power associated with the light emitted by the laser chain *before* the main pulse itself. In Fig. 8.5 the emitted power versus time is sketched in a log-log diagram. Though all the early emission is often indicated as *prepulse*, the actual prepulse (left hand peak in Fig. 8.5) is an ultrashort pulse, similar to the main pulse but much weaker, leaking from the electro-optical shutter out of the oscillator. This prepulse usually carries a negligible amount of energy (and power). More dangerous is the *amplified spontaneous emission* (ASE), also called *ASE pedestal*, which lasts typically a few nanosecond and then carries a considerable amount of energy, comparable with the main pulse energy if the contrast is worse than 10^6. In most of the previous experiments on laser-driven proton acceleration the contrast had to be increased to better than 10^9, with several means, including the "plasma mirror" technique [31, 37]. Early emission a few picosecond before the main pulse involves the *ps-contrast* which is usually 3–4 orders of magnitude worse than the ASE-contrast, but carries much less energy. This latter can be reduced only assuring high quality and accuracy in the optical compression of the stretched amplified pulse at the end of the laser chain. A critical feature of the pre-pulse problem is that most of the undesired effects depend on the absolute value of the pre-pulse energy and power and not from the value of the contrast. In other words, increasing the laser power, as necessary at increasing the ion energy, the contrast has to be increased correspondingly.

8.2.3 Laser Upgrading to Catch Suitable Ion Beam Energy

It has been clear for a long time that, differently from electrons (see next sections), proton sources driven by laser need not only high pulse peak intensity but also high energy per pulse. A pioneering experiment from Lawrence Livermore National Laboratory, demonstrated high current proton beams of several tens of MeV's [38] with PW laser pulses whose high contrast was assured by a plasma-mirror technique [37]. Preliminary investigations were performed in many laboratories with femtosecond and picosecond pulses of different power. These investigations were quite useful to assess the validity of various schemes achievable at the available laser fluence but they also evidenced that for getting kinetic energy and mean proton current suitable for clinical application, a general laser upgrading was necessary. Further, a decisive progress of laser technology towards higher peak power, higher contrast (see above), higher repetition rate has to be faced.

As an example, at Kansai Photon Science Institute of JAEA (Japan) an Advanced Beam Technology Division has been created with the purpose of designing a new-concept laser facility able to drive protons to 200 MeV kinetic energy. Previous experimental campaign 2004–2011 at KPSI/JAEA was quite successful. By increasing laser intensity on target from 10^{18} up to 10^{21} W/cm^2 the maximum proton energy increased progressively from 0.4 upto 40 MeV. This latter however was found to be also an energy limit due to the performances of the existing laser system J-KAREN [39].

The upgrading from J-KAREN to J-KAREN-P, which will deliver multi-PW power, is based on the original J-KAREN technology [40, 41] including OPCPA amplifiers and ultra-high contrast of the output pulse [42].

It is interesting to consider the expected effect of such a laser upgrading on the energy of protons which can be accelerated by focusing the laser pulse up to 10^{22} W/cm^2 with a contrast $>10^{11}$. Both TNSA and RPD acceleration schemes qualitatively converge towards expectation of proton energy above 200 MeV, also supported by numerical simulations performed for both extended and mass-limited thin foil targets.

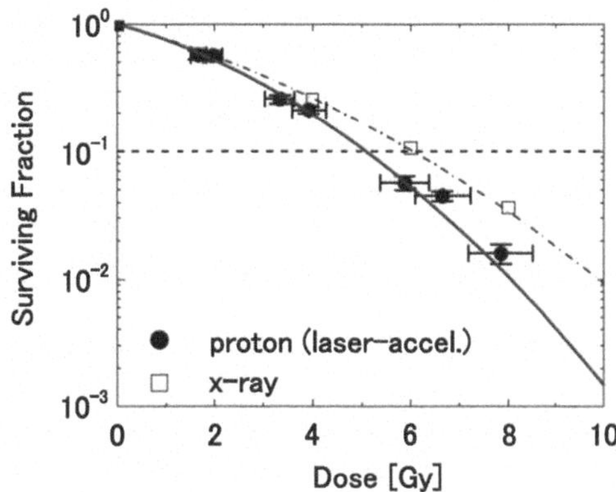

Fig. 8.6 Measured rate of survival of cancer cells after irradiation with laser produced protons (0.2 Gy/shot, LET = 18 keV/μm) with a determined RBE of 1.20 ± 0.11 (Yogo et al. [44])

KPSI/JAEA strategy for producing high energy protons is based on 40 fs laser pulses and consequent choice of acceleration schemes and target thickness and size. Some other labs have a different strategy according to the choice of longer (picosecond) or shorter (≈10 fs) laser pulses, respectively. Acceleration mechanisms and targetry will be varied consequently.

8.2.4 Radiobiology with Present Laser-Produced Ion Beams

Though ions produced with laser-plasma techniques are still far from clinical requirements, they are currently used for preliminary tests on biological samples in order to assess their capability as ionizing agent, also considering the ultra-short duration of the laser-produced particle bunches, compared with the ones delivered by RF-based machines. Relatively low kinetic energy, broad energy spectrum and large divergence of the beams do not prevent possibility and interest for such investigations.

Taking into account their high-LET (linear energy transfer), protons of a few MeV have been compared, in terms of relative biological effectiveness (RBE), with both RF-accelerated protons and standard X-ray sources. Yogo et al. have first demonstrated breaking of DNA in human cancerous cells with laser-accelerated protons [43], then measured their RBE [44] as shown in Fig. 8.6. A relevant feature of laser-produced proton bunches lies on their outstanding instantaneous dose rate, due to their duration of about 1 ps, more than one million times shorter than RF-produced pulses. Dose rate as high as 10^9 Gy/s have been obtained and tested on biological samples [45].

In this condition a single laser shot exposure could allow a clonogenic assay on tumor cells which appeared to be in line with previously published results employing RF-driven proton sources. On the other hand, Relative Biological Effectiveness (RBE) was estimated to be of the order of 1.4 at 10 % survival from a comparison with a "standard" 225 kVp X-ray source [45].

Ultra-high dose rate opens investigation of an unexplored regime of radiobiology, where collective effects [46] could add to individual particle effects on the cell, due to the extremely high particle density. On the other hand, ultra-short duration may open the way for studying (e.g. via pump/probe experiments) very early elementary processes occurring at a nanometer scale at the occurrence of the "instantaneous" ionizing action. These early physical and chemical processes are basically unknown. This exciting scientific perspective will be discussed also for ultrashort electron bunches (Sect. 4.2).

All these novel studies have to face a considerable difficulty due to the lack of suitable dosimetry for particle bunches of such a short duration and such a high dose rate. This point is common to laser-driven sources of both hadrons and electrons (Sect. 4.1). Novel devices and methods for dosimetry of ultrafast particle bunches are being proposed and tested recently. Fiorini et al. developed a new procedure [47] including the use of a magnet and Gafchromic films which, after a suitable Monte Carlo treatment of data, can provide both dose and energy spectrum of protons reaching the sample. This sophisticated procedure however needs the application of two correction factors: the first one, to take into account the variation of the dose response of the films as a function of the proton energy, and the other to calculate the dose at the cell layer from the dose measured on the films.

8.2.5 Laser-Driven Ion Microscopy

Ions produced with laser-based techniques can be used for high-resolution ion microscopy (ionography) of biological sample. For this application high kinetic energy of particles is not needed and also sub-MeV ions can produce high quality images provided the sample thickness is below but near the stopping range for the given energy band. It has to be considered that for contact microscopy also a high divergence of the ion beam can be tolerated, provided the spatial distribution of particle on the sample is uniform.

Faenov et al. [48] produced $>10^{18}$ multicharged carbon and oxygen ions per laser shot by irradiating CO_2 clusters from a gas-jet. The ion energy was measured to be $\geq$300 keV. With such rather divergent but uniform ion beam, ionography of a spider net revealed submicron details, as shown in Fig. 8.7.

8.3 Laser-Driven Electron Acceleration Towards Medical Applications

Starting from the "famous" triple communication of achievement of quasi-monochromatic electron bunches, by ultrashort intense laser interaction with gas-jets, 10 years ago [49–51], soon followed by the GeV achievement [52], an incredible crop of progressive successes has been reported in the scientific literature. Kinetic

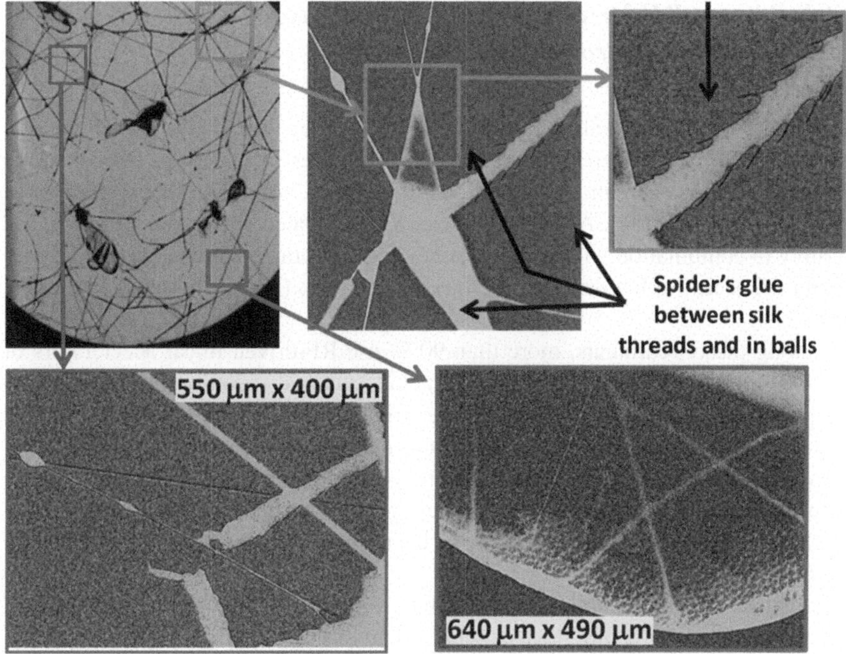

Fig. 8.7 Ion microscopy with carbon and oxygen ions produced by laser interaction with a CO_2 clustered jet: contact image of a spider web and magnified details showing high resolution (Faenov et al. [48])

energy and "quality" of the electron bunches have been enormously increased, in terms of energy control, spectrum narrowing, collimation, pointing stability, reproducibility of the process, by using a variety of targets, mostly gaseous, but also solids, these latter including thin dielectric foils. Recently electron bunches peaked at 2 GeV with only a few per cent energy spread and sub-milliradian divergence were reported [53] by focusing into a pulse-filled He cell a PW laser pulse. Beside these exciting records obtained with such powerful lasers, a great effort is devoted in many laser labs in order to obtain electron bunches suitable for medical applications, first of all radiotherapy of cancer. Since energy suitable to this kind of application ranges from a few up to a few tens of MeV, smaller laser equipment's are requested, typically a few tens of TW peak power. In principle, this kind of "table top" laser driven electron accelerators (Laser-Linac) are good candidates at competing with RF-based Linac currently used for radiotherapy in hospitals. Suitable kinetic energy and dose delivered in an acceptable treatment time are already available from Laser-Linac's. However, kinetic energy control, stability and reproducibility of the process have still to be improved in order to match clinical standards. On the other side, the much higher peak-dose-rate (of the order of 10^9 Gy/s) that can be achieved with laser techniques, opens an exciting field of radiobiological investigation with unpredictable consequences for the radiation therapy of tumors.

8.3.1 Present RF-Driven Versus Possible Laser-Driven Devices (Electrons and Photons)

Though the rate of survivals increases regularly year by year, cancer is still the first cause of death everywhere. The number of new cases of cancer in the world are estimated to have been about 13 millions in the year 2010, with an expectation of 20 millions in 2020 [54]. About 50 % of cases are treated with radiation therapies, possibly in combination with surgery and/or chemotherapy, with an emerging problem for the access of low- and middle-income countries (LMIC) to radiation therapy [55].

Among these treatments, more than 90 % use RF-driven linear accelerators of electrons (RF-Linac). Other techniques include internal radiation (brachytherapy) and proton-ion beams. These latter were discussed in the previous Sect. 8.2. In most cases electrons delivered by a RF-linac are not used directly on the tumor but converted into photons (hard X-rays) by bremsstrahlung through a suitable target. In some case electrons are used directly, either to cure superficial tumors or in the Intra-Operative Radiation Therapy (IORT) which can be applied during surgical operation of a tumor [56, 57].

Radiation therapy techniques evolve and progress continuously and so do accelerators and dose delivering devices which share a global market of $3–4 billions, growing at an annual rate of 5.4 % [58]. Most of the progress involves precision in tumor targeting, multi-beam irradiation, reduction of damage on healthy tissues and critical organs, fractionation of dose delivering for a more effective cure [59]. However, requirements on the electron beams to be suitable to a clinical use remain unchanged and they supply the benchmark for the progress of the laser-based clinical accelerators.

Basically, requested electron kinetic energy ranges from 4 to 25 MeV, but rarely energy above 15 MeV is used. Required dose/rate usually ranges from 1 to 10 Gy/min. These two ranges of performances are presently well fulfilled by plasma accelerators driven by ultrashort laser pulses of "moderate" peak power, i.e. few tens of TW, operating within high efficiency laser-plasma interaction regimes at a pulse repetition rate of the order of 10 Hz [60]. However further work has to be done on laser acceleration in order to reach the clinical standard in terms of the electron output stability and reproducibility.

Several tasks have to be afforded before proceeding to a technical design of a laser-driven linac prototype for clinical tests. A first task is the optimization of both laser and gas-jet (or other possible targets) as well as their coupling (involving mechanical stability and optical design). Another task is the energy control of the electron bunch to provide different electron energies on clinical demand. These goals would require a complex scientific and technological investigation addressed to both the laser system, in order to make it as stable, simple and easy to use as possible and to the physics of the acceleration process, in order to get the highest possible efficiency, stability and output control [61].

We may nevertheless try and list some of the expected advantages of future Laser-linac's for clinical uses. Laser technology strongly reduces size and complexity of the acceleration section (Mini-linac) of the device; it also totally decouples the "driver" from the acceleration section: we can imagine a single high power laser plant in a dedicated hospital room (with no need for radioprotection) which delivers pulses to a number of accelerators located in several treatment or operating rooms, suitably radioprotected. Laser managing and maintenance can proceed independently from the managing and maintenance of the Mini-linac's. Each Mini-linac could be easily translated and rotated according to the given radiotherapy plan. Current studies (see also Sects. 4.1 and 4.2) could prove that the extreme dose-rate per pulse delivered by the Laser-linac would reduce the total dose for a therapeutical effect. This latter of course would be a major advantage.

8.3.2 High Efficiency Laser-Driven Electron Acceleration to Relativistic Energies

If we limit our consideration to radiotherapy, present table-top laser driven electron accelerators can be already considered as candidate. In fact, for this medical application, most of the requirements usually asked to electron bunches are considerably relaxed. Small divergence, monochromaticity, pointing stability, etc. are requested at a moderate level, while the main effort has to be devoted to efficiency, stability and reliability of the process in order to provide clinically acceptable devices.

As far as the efficiency is concerned, in an experiment performed at CEA-Saclay (France) a regime of electron acceleration of high efficiency was found, using a 10 TW laser and a supersonic jet of Helium [60]. This *table-top* accelerator delivered high-charge (nC), reproducible, fairly collimated, and quasimonochromatic electron bunches, with peak energy in the range 10–45 MeV. In Fig. 8.8 a typical cross

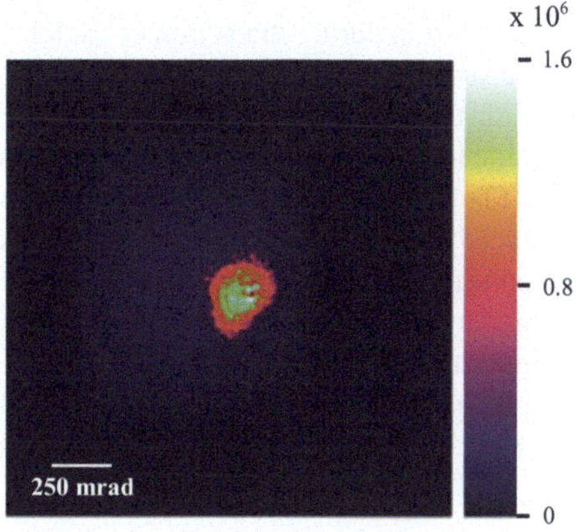

Fig. 8.8 25-MeV electron beam cross section (Giulietti et al. [60])

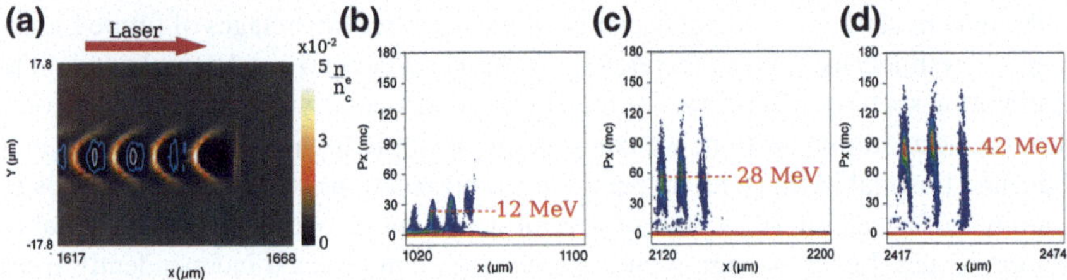

Fig. 8.9 Snapshots from the 3D simulation. **a** Electron density distribution when the laser pulse is at about 1.65 mm after the entrance in the gas jet. **b–d** Electron momentum distribution along the propagation axis when the laser pulse is at 1, 2, and 2.5 mm, respectively (Giulietti et al. [60])

Table 8.2 Comparison between commercial RF-linac's and experimental Laser-linac

Linac	IORT-NOVAC7	LIAC	Laser-linac (experimental)
Company	(SORDINA SpA)	(Info & Tech Srl)	(CEA-Saclay)
Max electron energy	10 MeV	12 MeV	45 MeV
Available energies	(3, 5, 7, 9 MeV)	(4, 6, 9, 12 MeV)	(5–45 MeV)
Peak current	1.5 mA	1.5 mA	>**1.6** KA
Bunch duration	4 μs	1.2 μs	<**1** ps
Bunch charge	6 nC	1.8 nC	1.6 nC
Repetition rate	5 Hz	5–20 Hz	**10** Hz
Mean current	30 nA @5 Hz	18 nA @10 Hz	**16** nA @10 Hz
Released en. in 1 min	18 J @ 9 MeV	14 J @12 MeV	**21** J @20 MeV

section of the relativistic electron beam at 25 MeV is shown, after de-convolution of experimental data from the SHEEBA radiochromic film stack device [62].

3D particle-in-cell simulation performed with the numerical code CALDER [63] reveals that the unprecedented efficiency of this accelerator was due to the achievement of a physical regime in which multiple electron bunches are accelerated in the gas-jet plasma during the action of each laser shot. This effect is shown in Fig. 8.9 by the simulation sequence.

With this experiment, laser driven electron acceleration approached the stage of suitability for medical uses, in particular for Intra-Operative Radiation Therapy (IORT) of tumors [56, 57]. Comparison of the main parameters of electron bunches produced by a commercial RF Hospital accelerator for IORT treatment and those of the present laser driven accelerator is shown in the Table 8.2.

In the same experiment electron bunches of ≈40 MeV were converted, via bremsstrahlung in a tantalum foil, into gamma rays with a strong component in the range 10–20 Mev, which matches the Giant Dipole Resonance of nuclei. This gamma rays could in turn activate a foil of gold according to the nuclear reaction

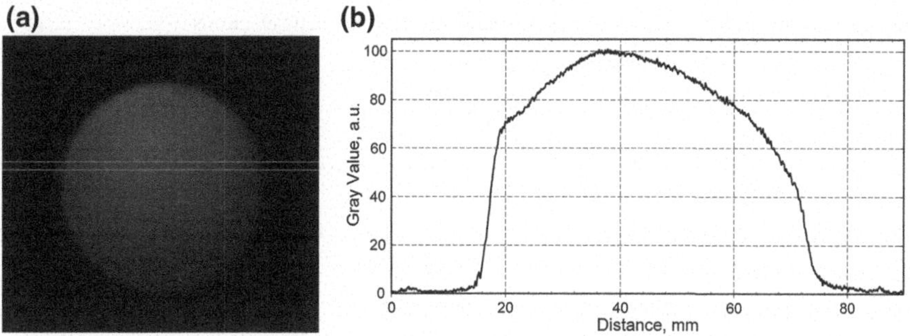

Fig. 8.10 Distribution on sample of electrons from the 300 keV, high-RBE electron source

^{197}Au(γ, n)^{196}Au. The number of radioactive gold atoms produced in this way was measured [1, 60]. This achievement opens the way to table-top laser-driven nuclear physics and production of radio-isotopes for medical uses.

8.3.3 Sub-relativistic Electron Sources

Recently, a laser-driven source of electron bunches with kinetic energy around 300 keV and picosecond duration has been set up at ILIL lab of National Institute of Optics (Pisa, Italy) for radiobiological tests. Each bunch combines high charge with short duration and sub-millimeter range into a record dose rate, exceeding 10^9 Gy/sec. Both high dose rate and the high level of Relative Biological Effectiveness (RBE), attached to sub-MeV electrons, make this source very attractive for radiobiological tests on thin living samples. The source reliability is improved by its shot-to-shot stability and uniform dose distribution on the sample surface, this latter shown in Fig. 8.10. Operating at 10 Hz, the source can deliver high doses in a short exposure time. The physical and radiological performances of the source are indicated in the Table 8.3.

Preliminary tests on biological samples confirm the high performances of the source [64] and its potential for radiobiological studies. In fact, epidemiological and experimental analysis indicates that low-LET (Linear Energy Transfer) radiations (X-rays, electrons) have different RBE's depending on their energy. In particular, higher-energy low-LET radiations are relatively less effective than lower-energy low-LET radiations. In particular, there is a factor of about 3–6 increase in effectiveness in 30 kVp X-rays and for tritium beta rays compared to 15 MeV electrons [65]. On the other hand the limit that low energy radiation is absorbed in a small penetration depth turns out in an increased dose for thin bio-samples.

Table 8.3 Main performances of the ILIL source of sub-relativistic electrons

Number of electrons N_e per bunch	$N_e \approx 10^{10}$
Bunch duration on sample	$\approx$3.5 ps
Bunch length L	L $\approx$ 1mm
Kinetic energy (E) distribution	Approx. Maxwellian $\approx \exp[-E/kT]$
Source temperature T	T$\approx$300 keV
Stopping power in water (E = 300 keV)	2.36 MeVcm^2 /g
Range in water (E = 300 keV)	0.84 mm
Dose released by each bunch in water	3.5 mGy
Dose rate in water during the bunch	10^9 Gy/s
Bunch repetition rate	10 Hz
Multi-bunch dose rate	35 mG/s
Time for delivering 1 Gy	$\approx$30 s

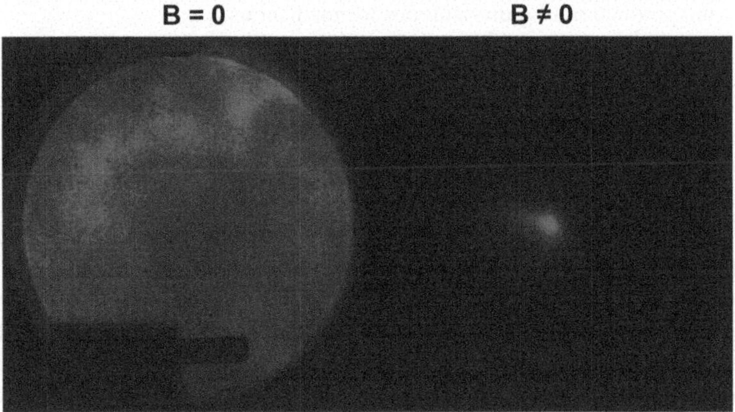

Fig. 8.11 Collimation effect of external magnetic field

Such sub-relativistic electron beams can be also easily focused by an external magnetic field in case the irradiation has to be confined in a small area of the sample. Recently, several magnetic field configurations have been successfully tested in order to find the optimum focusing on sample for an electron beam of 300 keV [66]. Figure 8.11 clearly shows that, with a suitable magnetic configuration, even an initially non-uniform electron beam can be strongly focused on the sample position. The same work also proved that the magnetic field can be beneficial also in stabilizing the electron beam pointing on a given target spot.

8.4 Radiological Use of Electron Bunches from Laser-Plasmas

8.4.1 Ionizing Electron Bunches for Biological Tests and Advanced Dosimetry

As of today, the basic ionizing radiation used in cancer radiotherapy is represented by X-rays and electron beams in the energy range of 4–25 MeV. Nevertheless, new irradiation sources based on laser-plasma accelerated electrons are now emerging, characterized by ultra-short particle bunches (∼ps) and ultra-high dose rate that could result in peculiar radiobiological properties [67]. In order to be actively used in the future for cancer radiotherapy, laser-plasma electron sources have to meet requirements as efficiency, long-term stability, reproducibility and reliability of the process in order to provide clinically acceptable devices. Moreover, sufficient particle intensities and controlled delivery of the prescribed dose at the treatment site has to be assessed [68].

Radiological use of laser-plasma accelerated electrons implies a well-defined dosimetric characterization. Apart from retrospective precise dose determination and dose homogeneity control by means of Faraday cup and GafChromic films, Monte Carlo simulations performed with the code GEANT4 have allowed the "a priori" characterization of dosimetric properties of relativistic electron beams produced in laser-plasma acceleration. Figure 8.12 shows the simulated dose deposited by a 6 MeV electron bunch at various depths inside a water phantom [69].

Following the translational research chain from bench to bedside, the laser based technology was developed for cell irradiation experiments as the first translational step. Its stable and reliable application was proven in systematic radiobiological studies like the one performed at Friedrich Schiller University with the JETI laser system [70]. Tumor and non-malignant cells were irradiated with pulsed laser-accelerated electrons in the range 3–20 MeV for the comparison with electrons of a conventional Linac for therapy. Dose response curves were measured for the biological endpoints, clonogenic survival and residual DNA double strand breaks. The overall results show no significant differences in radiobiological response for in vitro cell experiments

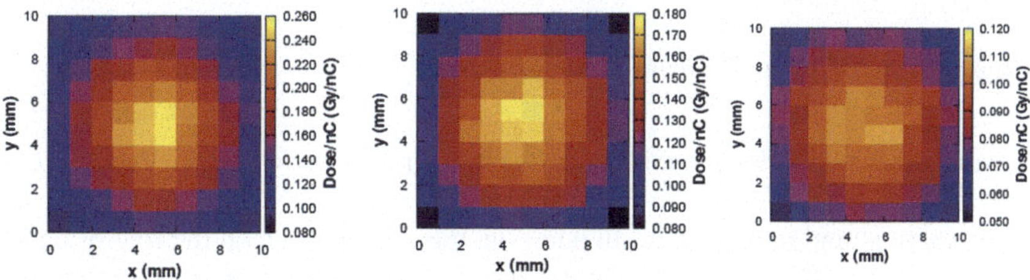

Fig. 8.12 2D maps of the dose deposited by e− at 0.5, 4.5 and 9.5 mm depth (Labate et al. [69])

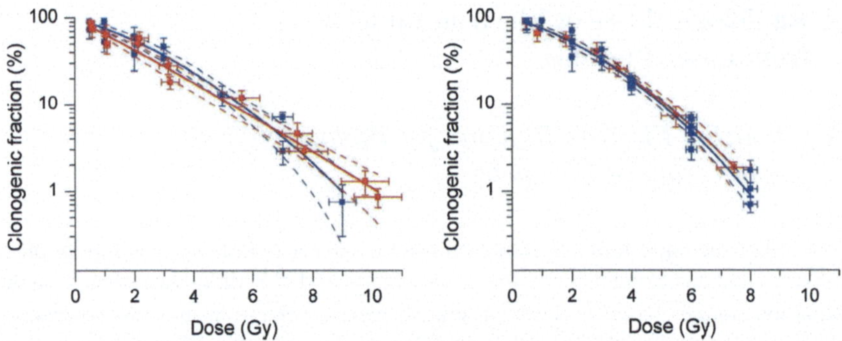

Fig. 8.13 Dose response curves of clonogenic survival after irradiation of normal tissue cell line 184A1 (*left*) as well as tumor cell line FaDu (*right*) with pulsed laser generated electron beam (*red curves*) and continuous electron beam delivered by a therapeutic LINAC (*blue curves*). All curves are presented in consideration of their 95 % confidence intervals (Laschinsky et al. [70])

between laser and RF produced electron beams, suggesting that no systematic difference exists between the two sources. Figure 8.13 shows dose response curves of clonogenic survival after irradiation with pulsed laser generated electron beam and continuous electron beam delivered by a therapeutic LINAC.

Similar experiments have been conducted on cell monolayer and lymphocytes at ILIL-INO laboratory of the National Research Council in Pisa, using a 2 TW Ti:Sa laser and accelerated electron in the range 5–20 MeV. The electron bunches have been fully characterized, in terms of spectrum, divergence and total charge, using GAFChromic Films (GAF) packed in a stack and separated by solid water, and GEANT4 library has been used to retrieve the dose delivered to the biological samples. DNA double-strand breaks (DSB) and nucleo-cytoplasmic translocation alterations have been inspected as biological endpoint [71].

Animal experiments have been carried out at Friedrich Schiller University, proving the feasibility of a laser based irradiation system with all key components as beam transport system, real-time beam monitoring, absolute dosimetry and reproducible positioning of the tumor on beam axis [72]. However, new insights into completely unexplored domains as cell and tissue responses to pulsed ultra-high dosed rates may also be provided by ultra-fast spatio-temporal radiation biology and dosimetry.

8.4.2 "Femto-Nano" Scale Radiobiology

Conventional radiobiology studies and measures the effects of ionizing radiation on living cells, mostly considering cell death and apoptosis, DNA damage and repair, recombination, and mutagenesis. In addition, there is research on the role of oncogenes in cancer, RNA processing in eukaryotes and radiobiology topics related to cancer therapy. All these observations and measurements usually lie on a time scale ranging from few minutes to several days.

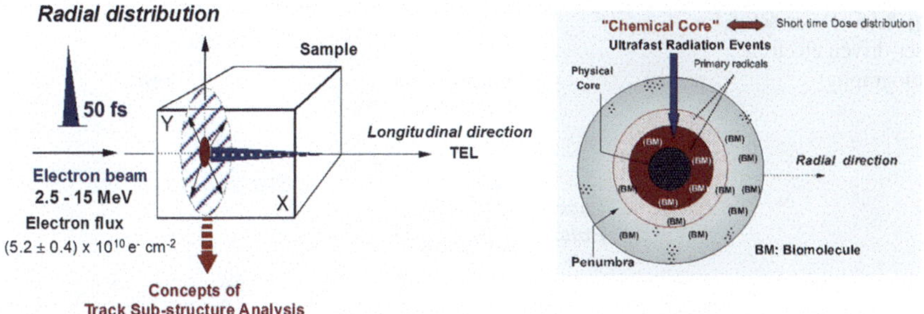

Fig. 8.14 Basic layout of High Energy Radiation Femtochemistry (HERF) concept. Courtesy of Y. Gauduel

Early deposition of energy and energy spreading in a cascade process is usually described by Monte Carlo code simulations but very few is known about the "instantaneous" response of the components of a living cell at this early stage. On the other hand, the early profile of energy deposition is decisive for the prediction and control of radiation-induced biomolecular and sub-cellular damage, and consequently for cancer therapy protocols. This point could become of extreme relevance if ultra-short, laser driven, ionizing pulses of radiation/particles will be used in the future radiotherapy.

That's why recently a novel approach to radiation biology has been proposed [73, 74] for the complete understanding of biophysical events triggered by an initial energy deposition inside confined ionization clusters (tracks) and evolving over several orders of magnitude, typically from femtosecond and sub-nanometer scales up to the scale of conventional radiobiology. To this purpose, femtosecond laser sources providing ultra-short pulses of both optical photons and relativistic electrons, in the eV and MeV domain respectively, open exciting opportunities for a real-time imaging of radiation-induced biomolecular alterations in nanoscopic tracks. Figure 8.14 shows the basic spatio-temporal geometry of this approach.

Recently, using a very short-lived quantum probe (2p-like excited electron) and high-time resolved laser spectroscopic methods in the near IR and the temporal window 500–5000 fs, Gauduel et al. could demonstrate that short-range coherent interactions between the quantum probe and a small biosensor of 20 atoms (disulfide molecule) are characterized by an effective reaction radius of $\approx$1 nm. For the first time, femtobioradical investigations performed with aqueous environments gave correlated information on spatial and temporal biomolecular damages triggered by a very short lived quantum scalpel whose gyration radius is around 0.6 nm [16].

There is the hope that this innovating approach would be applied to more complex biological architectures such as nucleosomes, healthy and tumor cells. In the framework of high-quality ultra-short penetrating radiation beams devoted to pulsed radiotherapy of cancers, this concept would foreshadow the development of real-time nanobiodosimetry combined to highly-selective targeted pro-drug activation.

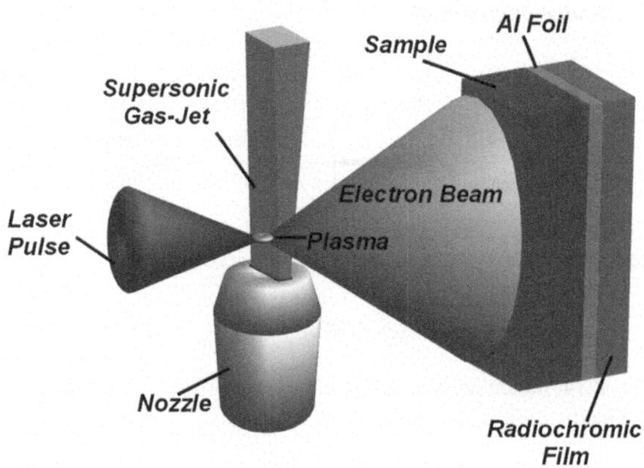

Fig. 8.15 Basic setup for laser-driven electron radiography

8.4.3 Laser-Driven Electron Radiography

Techniques for innovative imaging using electrons represent an important non destructive tool for material science, biology and medicine [75]. Though the use of the laser plasma based electron accelerators for electron radiography is only at the beginning, it can be considered one of the most promising approaches for good quality imaging with particle beams obtainable with miniaturized equipment compared with standard accelerators [73]. Presently, there are only few examples of application of this technique in the field of particle radiography, one of which with laser-produced ions as described in Sect. 2.4 above [48], another with electrons, but generated with conventional linear accelerators [76]. Actually, laser-based accelerators can produce electron beams with enough energy that can penetrate also in dense material [75] and they are certainly an attractive source for probing solid objects on a large area (several squared centimeters). Moreover the compactness of table-top laser accelerators, compared with conventional accelerators, makes this kind of radiography much easier.

The setup is typical of a laser-plasma accelerator: a multi-TW femtosecond laser beam is focused on a gas jet, producing high energy electrons, as shown in Fig. 8.15. Particularly effective is the use of clustered gas jet [77, 78]. In the following example [79] high pressure Argon was released in the pulsed gas-jet, in condition of strong clusterization. The laser was a 2-TW Ti:Sapphire. In this way high charge electron bunches were obtained with a uniform distribution of electrons on the sample, similarly the one shown in Fig. 8.12. About 30 laser shots were enough to obtain a good quality electron radiography.

A typical sample (about 8×5 cm) is shown in Fig. 8.16 left side. In particular this sample included both inorganic and organic elements to test the capabilities of the system in generating radiography. The samples are placed at about 10 cm from the gas jet and, just after the sample, in direct contact, two radiochromic dosimetry films are inserted (Fig. 8.15).

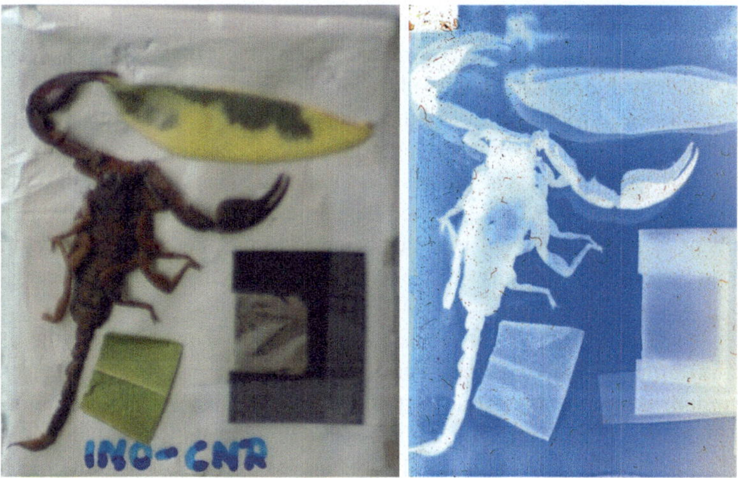

Fig. 8.16 The sample (*left*) and its own electron radiograph. *Overall size* 35 mm x 48 mm (Bussolino et al. [77])

The possibility of radiographic images generation is proved in Fig. 8.16 right hand side. The image shows detailed features for all elements (inorganic and organic, thin and thick) inserted in the sample, with high contrast. The resolution of the radiography shown in Fig. 8.16 has been estimated to be ≤ 60 μm, higher than the one obtained in previous works [76, 80]. Simple optical calculations show that with this setup the resolution can be further improved down to 10 μm [79].

Transmission electron radiography of organic and inorganic dense objects over a field of view more than 50 mm wide could be easily achieved. The images show details of both thicker and thinner features. The spatial resolution in the current geometrical configuration was limited by geometrical effects combined with the intrinsic detector resolution and scattering from the sample. These results suggest that the availability of a new laser-driven electron source could be taken into account as a substitute to conventional RF electron guns for a coste ffective approach to high resolution transmission electron microscopy at high electron energy and for configurations where multiple plasma sources are needed.

8.5 Conclusions

Ultra-short intense laser science has been providing a new class of plasma sources which deliver high energy particles. These sources are candidate at competing with the existing Radio-Frequency driven particle accelerators for many applications, including radiography, radiobiology, radiotherapy and nuclear medicine. Unique features of these novel sources are the reduced size of the accelerating device and the much shorter duration of the particle bunches. This latter feature can allow ultrafast imaging in radiology, investigation of very early precursors of the biological response and ultra-high peak dose rate in the radiological cure of cancer.

As long as the achievement of clinical standard is concerned, the status of the art is quite different for the laser-driven proton/ion acceleration respect to electron acceleration. Laser-plasma accelerators can easily produce electrons of the requested kinetic energy and deliver the requested dose in a typical treatment time. However they still need to be improved in order to provide the requested electron energy control, as well as the necessary degree of stability and reliability of the acceleration process.

Laser-based proton and light ion acceleration did not reach so far the clinical request in terms of kinetic energy, the necessarily narrow energy spectrum nor the minimum dose rate. Nevertheless the technique is expected to face a decisive improvement when the current upgrading of the laser systems devoted to it will be completed. Each of these upgrading effort is aimed to match a particular acceleration regime among several promising regimes designed by theoretical and numerical calculations. To this respect, the physical investigation is quite active due to the high value of any progress towards the clinical application. In fact, if successful, laser-driven ion accelerators could widely expand the number of hadron-therapy treatments, presently limited by the size and cost of RF-based infrastructures.

At meantime a large crop of data is being currently collected from radiobiological experiments performed with both electron and ion sources based on laser-plasmas on living samples. This novel and challenging investigation will provide the necessary background for the future use of laser based clinical devices in terms of (presently unknown) biological response and ad hoc dosimetry.

Acknowledgments The authors of this Chapter are operating in the framework of the CNR High Field Photonics Unit (MD.P03.034). They acknowledge financial support from the CNR funded Italian research Network 'ELI-Italy (Attoseconds)', from the Italian Ministry of Health funded project GR-2009-1608935 (D.I. AgeNaS) and from the INFN funded "G-RESIST" project.

References

1. O. Graydon, Nature Photonics 7, 585 (2013), A. Giulietti and A. Gamucci, Progress in Ultrafast Intense Laser Science, Vol. V, Ch. 8, Springer Series in Chemical Physics (Springer, Heidelberg, 2010)
2. G.A. Mourou, T. Tajima, S. Bulanov, Rev. Modern Phys. **78**, 309 (2006)
3. S.P. Le Blanc, R. Sauerbrey, S.C. Rae, K. Burnett, J. Opt. Soc. Am. B **10**, 1801 (1993)
4. D. Giulietti et al., Phys. Rev. Lett. **79**, 3194 (1997)
5. J.K. Koga, N. Naumova, M. Kando, L.N. Tsintsadtze, K. Nakajima, S.V. Bulanov, H. Dewa, H. Kotaki, T. Tajima, Phys. Plasmas **7**, 5223 (2000)
6. A. Giulietti, P. Tomassini, M. Galimberti, D. Giulietti, L.A. Gizzi, P. Koester, L. Labate, T. Ceccotti, P. D'Oliveira, T. Auguste, P. Monot, Ph Martin, Phys. Plasmas **13**, 093103 (2006)
7. A. Giulietti, A. André, S. Dobosz, Dufrénoy, D. Giulietti, T. Hosokai, P. Koester, H. Kotaki, L. Labate, T. Levato, R. Nuter, N. C. Pathak, P. Monot, and L. A. Gizzi. Phys. Plasmas **20**, 082307 (2013)

8. T. Tajima, J. Dawson, Phys. Rev. Lett. **43**, 267 (1979)
9. D. Strickland, G. Mourou, Opt. Commun. **56**, 219 (1985)
10. W. Priedhorsky, D. Lier, R. Day, D. Gerke, Phys. Rev. Lett. **47**, 1661 (1981)
11. D.M. Villeneuve, G.D. Enright, M.C. Richardson, Phys. Rev. A **27**, 2656 (1983)
12. V. I. Veksler, in *Proceedings of CERN Symposium on High Energy Accelerators and Pion Physics*, vol. 1, p. 80 (Geneva, Switzerland, 1956)
13. A.P. Fews, P.A. Norreys, F.N. Beg, A.R. Bel, A.R. Dangor, C.N. Danson, P. Lee, S.J. Rose, Phys. Rev. Lett. **73**, 1801 (1994)
14. S. Bulanov, T. Esirkepov, V. Khoroshkov, A. Kuznetsov, F. Pegoraro, Phys. Lett. A **299**, 240 (2002)
15. S.S. Bulanov et al., Med. Phys. **35**, 1770 (2008)
16. Y.A. Gauduel, V. Malka, *Proceedings of SPIE* 8954; doi:10.1117/12.2038983 (2014)
17. V. Malka et al., Med. Phys. **31**, 1587 (2004)
18. http://www.cnao.it/index.php/en/
19. http://neurosurgery.mgh.harvard.edu/protonbeam/nptcbrochure.pdf
20. R.R. Wilson, Radiology **47**, 487 (1946)
21. J. Lawrence, Cancer **10**, 795 (1957)
22. S. Sawada, Nucl. Phys. A **834**, 701 (2010)
23. M. Goitein, A.J. Lomax, E.S. Pedroni, Phys. Today **55**, 45 (2012)
24. M. Schippers, *Beam Delivery System for Particle Therapy*, in *Proton and Ion Carbon Therapy*, C.-M. Charlie Ma, T. Lomax (eds.), CRC Press (Boca Raton, FL) p. 43 (2013)
25. P. Mulser and D. Bauer, *High Power Laser-Matter Interaction*, Springer Tracts in Modern Physics, vol. 238 (Springer, New York, 2010)
26. A. Macchi, M. Borghesi, M. Passoni, Rev. Mod. Phys. **85**, 751 (2013)
27. H. Daido, M. Nishiuchi, A.S. Pirozhkov, Rep. Prog. Phys. **75**, 056401 (2012)
28. S.C. Wilks et al., Phys. Plasmas **8**, 542 (2001)
29. Mackinnon et al., Phys. Rev. Lett. **86**(1769) (2001)
30. Zeil et al., New J. Phys. **12**(045015) (2010)
31. T. Ceccotti et al., Phys. Rev. Lett. **99**, 185002 (2007)
32. M. Kaluza et al., Phys. Rev. Lett. **93**, 045003 (2004)
33. T. Ceccotti et al., Phys. Rev. Lett. **111**, 18501 (2013)
34. S. Sinigardi et al., Nucl. Instr. Meth. Phys. Res. **A740**, 99 (2014)
35. L. Landau et E. Lifchitz, *Théorie du Champ,* Editions (MIR, Moscou, 1966)
36. M. Tamburini et al., New J. Phys. **12**, 123005 (2010)
37. M.D. Perry et al., Opt. Lett. **24**, 160 (1999)
38. R.A. Snavely et al., Phys. Rev. Lett. **85**, 2945 (2000)
39. M. Aoyama et al., Opt. Lett. **28**, 1594 (2003)
40. H. Kiriyama et al., Opt. Comm. **282**, 625 (2009)
41. H. Kiriyama et al., Opt. Lett. **35**, 1497 (2010)
42. K. Ogura et al., Opt. Lett. **37**, 2868 (2012)
43. A. Yogo et al., Appl. Phys. Lett. **94**, 181502 (2009)
44. A. Yogo et al., Appl. Phys. Lett. **98**, 053701 (2011)
45. D. Doria et al., AIP Adv. **2**, 011209 (2012)
46. E. Fourkal et al., Phys. Med. Biol. **56**, 3123 (2011)
47. F. Fiorini et al., Phys. Med. Biol. **56**, 6969 (2011)
48. A.Y. Faenov et al., Appl. Phys. Lett. **95**, 101107 (2009)
49. P.D. Mangles et al., Nature **431**, 535 (2004)
50. G.C.R. Geddes et al., Nature **431**, 538 (2004)
51. J. Faure et al., Nature **431**, 541 (2004)
52. W. Leemans et al., Nat. Phys. **2**, 696 (2006)
53. X. Wang et al., Nature Commun. **4**, Article no. 1988 (2013) doi:10.1038/ncomms2988
54. IARC, Cancer Fact Sheets, Globocan, http://globocan.iarc.fr/
55. D. Rodin et al., Lancet Oncol. **15**, 378 (2014)
56. U. Veronesi et al., Ann. Oncol. **12**, 997 (2001)

57. A.S. Beddar et al., Med. Phys. **33**, 1476 (2006)
58. http://www.businesswire.com/news/home/20140313005577/en/Research-Markets-External-Beam-Radiation-Therapy-Devices
59. R. Baskar et al., Int. J. Med. Sci. **9**, 193 (2012)
60. A. Giulietti et al., Phys. Rev. Lett. **101**, 105002 (2008)
61. F. Baffigi et al., The LEARC Concept: Laser-driven Electron Accelerator for Radiotherapy of Cancer, INO-CNR Internal Report (2014). Accessed: ilil.ino.it
62. M. Galimberti et al., Rev. Sci. Instrum. **76**, 053303 (2005)
63. E. Lefebvre et al., Nucl. Fus. **43**, 629 (2003)
64. L. Fulgentini et al., High RBE doses delivered by a laser driven electron source , INO-CNR Internal Report (2014). Accessed: ilil.ino.it
65. N. Hunter, C.R. Muirhead, J. Radiol. Prot. **29**, 5 (2009)
66. Y. Oishi et al., Jpn. J. Appl. Phys. **53**, 092702 (2014)
67. V. Malka, J. Faure, Y.A. Gauduel, Mutat. Res. **704**, 142 (2010)
68. E. Beyreuther et al., Med. Phys. **37**, 1392 (2010)
69. L. Labate et al, INO-CNR Internal Report (prot. 151, 13/01/11) (2011)
70. L. Laschinsky et al., J. Radiat. Res. **53**, 395 (2012)
71. L. Labate et al., Proc.SPIE **8779** (2013)
72. M. Schurer et al., Biomed. Tech. **57**, 62–65 (2012)
73. Y.A. Gauduel et al., Nat. Phys. **4**, 447 (2008)
74. Y.A. Gauduel et al., Eur. Phys. J. D **4**, 121 (2010)
75. S.P.D. Mangles et al., Laser Part. Beams **24**, 185–190 (2006)
76. F. Merrill et al., Electron radiography. Nucl. Instrum. Method Phys. Res. **B261**, 382–386 (2007)
77. A.Ya. Faenov et al., Laser Part. Beams **26**, 69 (2008)
78. P. Koester et al., Laser Part. Beams **33**, 331 (2015)
79. G.C. Bussolino et al., J. Phys D: Appl. Phys. **46**, 245501 (2013)
80. V. Ramanathan et al., Phys. Rev. Special Topics Accel. Beams **13**, 104701 (2010)

Chapter 9
Observation of Ultrafast Photoinduced Dynamics in Strongly Correlated Organic Materials

Ken Onda

Abstract Complex photoinduced dynamics in organic charge transfer complexes, which have strong electron correlation and low-dimensional electronic systems, have been investigated using various ultrafast analytical techniques. A short-lived hidden photoinduced phase was determined from the transient reflectivity spectrum using a 120 fs pulse over a wide photon energy range from infrared to visible. The initial process of the photoinduced dynamics and its electronic coherence were revealed using a pulse compression technique down to 12 fs. The molecular and lattice structures were found to change in a different way from the electronic states, based on observations using two types of time-resolved vibrational spectroscopy and femtosecond electron diffraction. Time-resolved photoemission electron microscopy was developed for observing transient domain structures during photoinduced phase transitions.

9.1 Introduction

Phenomena induced by photo-irradiation in solid materials essentially differ from those induced by temperature increases. Upon temperature increase, electrons and phonons retain their statistical distribution, either Fermi or Bose distribution, and the theory of thermodynamics holds well. In contrast, photo-irradiation excites electrons and phonons directly to high energy levels and there is no comprehensive theory to describe this system far from thermal equilibrium. To understand complex phenomena in such non-equilibrium states, various ultrafast laser techniques have been developed, which operate in the range of femtoseconds to nanoseconds.

Strongly correlated materials have attracted considerable interest recently. Strong electron–electron interactions in these materials induce a wide variety of physical properties such as high-Tc superconductivity and gigantic magnetic resistance in

---

K. Onda (✉)
PRESTO, Japan Science and Technology Agency and Graduate School of Science and Engineering, Tokyo Institute of Technology, S1-8, 4259 Nagatsuta, Midori-ku, Yokohama 226-8502, Japan
e-mail: onda.k.aa@m.titech.ac.jp

K. Yamanouchi et al. (eds.), *Progress in Ultrafast Intense Laser Science XII*,
Springer Series in Chemical Physics 112, DOI 10.1007/978-3-319-23657-5_9

thermal equilibrium [1]. In non-equilibrium states created by photo-irradiation, these materials are also expected to exhibit several new physical properties. One intriguing phenomenon expected in these materials is photo-induced phase transition (PIPT) [2, 3]. In this phenomenon, a few photons can trigger a change in the macroscopic physical properties cooperatively through strong electron–electron and electron–lattice interactions, and it is expected to have application in ultrafast pure optical switching and efficient photo-energy conversion. However, the ultrafast speed and complication of the process prevents the study of PIPT using conventional ultrafast spectroscopic techniques. In this review, we take charge transfer complexes as a typical example of organic strongly correlated materials and introduce our ultrafast techniques to study the ultrafast photoinduced dynamics including PIPT, and consider how the obtained data can be interpreted. The techniques shown here are widely applicable to studies of photoinduced dynamics in other types of solid materials.

Charge transfer (CT) complexes are organic single crystals consisting of two types of π-conjugated small molecules [4, 5]. One of the molecules tends to donate an electron, i.e. is a donor molecule, and the other tends to accept an electron, i.e. is an acceptor molecule. Thus, electron transfer naturally takes place when they form a crystal and an unfilled band is formed in the crystal. Because of this phenomenon, CT complexes possess metal- or semiconductor-like characters despite being organic materials. The small overlap between the π-orbitals of constituent molecules (the transfer integral, $t \sim 0.1\,\text{eV}$) increases the interaction between electrons and thus these complexes are regarded as strongly correlated electron systems. They also acquire unique physical properties originating from the low-dimensionality of their electronic structure owing to the anisotropic π-orbitals. Moreover, the weak interaction between constituent molecules causes the complexes to be flexible and soft. They are therefore affected significantly by ambient pressure, temperature, and electric fields. These unique characteristics, that is, strong electron correlation, low dimensionality, and soft structure, give rise to a wide variety of physical properties in thermal equilibrium, and photo-irradiation is expected to allow ultrafast control of these physical properties.

9.2 Overall Process of PIPT and Observation Techniques

Figure 9.1 shows the overall process of the photoinduced dynamics in strongly correlated materials along with the observation methods for each part of the process. Photo-irradiation excites the sample to the Franck–Condon (F-C) state, where only electrons are excited and the lattice structure is the same as the ground state. In this state, electronic coherence is induced by electric field oscillation of the excitation light, and in some cases strongly coupled phonons accompany this electronic coherence. The coherence decays within a few tens of femtoseconds due to interactions with other electrons or the lattice. To observe this ultrafast process, a pulse compression technique down to $\sim$10 fs is required (Sect. 9.4). Immediately after this, charge melting or redistribution takes place and the created state strongly depends

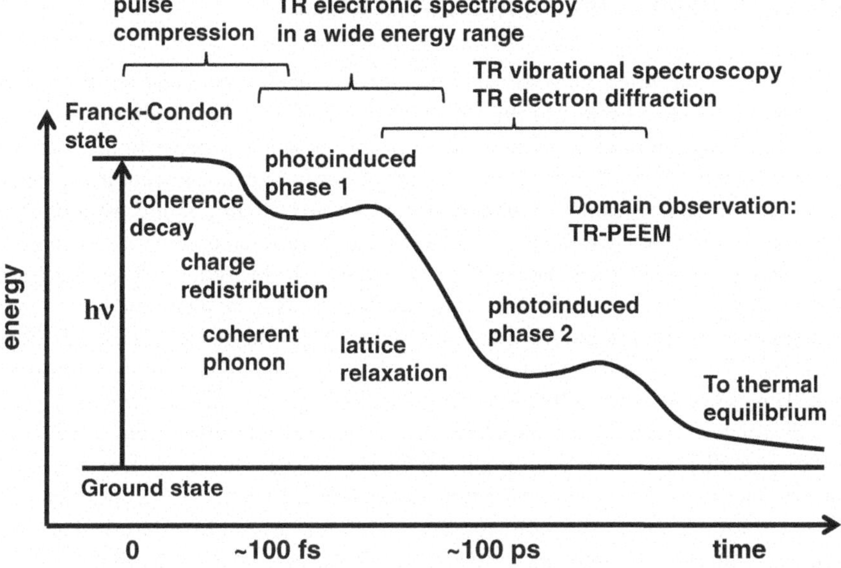

Fig. 9.1 Schematic summary of photoinduced dynamics and their observation methods in strongly correlated organic crystals

on electron–electron and electron–phonon interactions in the material. This non-equilibrium state is often considered to have a different physical property than that in the ground state and so it is called the "photoinduced phase" or "hidden phase". At around this time, the lattice is often excited coherently and is called a "coherent phonon". To assign this photoinduced phase, the transient electronic spectrum in a wide photon range from infrared to visible has to be measured (Sect. 9.3). After this highly non-equilibrium state, another state emerges, which is often similar to the high temperature phase. The time required for this variation is 10 fs to 100 ps depending on the material. At this stage, it is essential to determine the lattice and molecular structures as well as the electronic structure and thus structure sensitive methods such as time-resolved vibrational spectroscopy and electron diffraction are important (Sect. 9.5). In addition to these time-resolved methods, sub-micrometer-scale domains are expected to play an important role in PIPT, though they have never been directly observed, and therefore a method to observe spatial images on the nanometer spatial and femtosecond temporal scales has to be developed (Sect. 9.6). Finally, the system reverts to the state before photoexcitation within 1 ms, so that repeat measurements using pulse lasers can be made.

9.3 Assignment of the Photoinduced Phase

Most fundamental procedures to study the photoinduced process assign the short-lived electronic state by measuring its transient electronic spectra after photoexcitation. In solid materials, characteristic electronic transitions occur in the infrared region less than 1.5 eV. CT transitions between atoms or molecules occur in the range from 0.1 to 1.5 eV and the conduction electrons reflect light ranging from plasma frequency (ω_p), which is normally located in the near infrared region, to zero photon energy as expressed by the Drude model. To assign such electronic states, a spectrum in a wide photon energy range in the infrared region has to be measured.

In general, to understand the electronic states of a solid material, its reflectivity spectrum is measured, though this is only indirectly concerned with the electronic state. This is because it is difficult to obtain the absorption spectrum, which is more directly related to the electronic state, due to strong light absorption, especially in the infrared region, and the sample thickness. Using a Kramers–Kronig (K-K) transformation, the absorption spectrum and optical conductivity spectrum can be derived from the reflectivity spectra [6]. Spectral simulation based on physical models is another way to obtain information on electronic states from the reflectivity spectra. The Drude model, which assumes free electrons, and the Lorentz model, which assumes bound electrons, are often used to reproduce the reflectivity spectra [6].

Transient reflectivity spectra are measured using the pump–probe method. Figure 9.2 shows a typical experimental setup for this measurement on the 100 fs

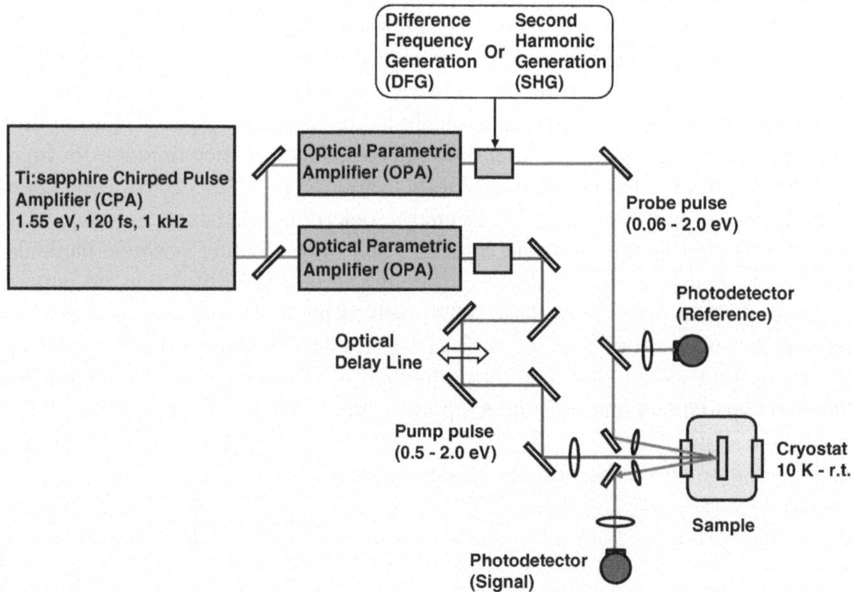

Fig. 9.2 Typical experimental setup for studying ultrafast dynamics in solid materials

time scale. The light source of this measurement is a conventional Ti:sapphire chirped pulse amplifier (CPA) with pulse width ∼120 fs, photon energy ∼1.55 eV (800 nm), pulse energy ∼1 mJ/pulse, and repetition rate ∼1 kHz. A tunable near-infrared probe pulse (0.5–1.0 eV) is obtained from the signal and idler of the optical parametric amplification (OPA) of the CPA. A mid-infrared pulse is generated by difference frequency generation (DFG) between the signal and idler using a AgGaS crystal (0.12–0.5 eV) or a GaSe crystal (0.06–0.12 eV). A tunable visible pulse is obtained by second harmonic generation (SHG) of the signal (1.0–2.0 eV). A pump pulse is often obtained from the output of the CPA, and sometimes from the signal and idler of the OPA because the charge transfer transition that triggers PIPT is generally located in the near infrared region [3]. In addition to these pump and probe pulses, we must select proper optics, such as mirrors, detectors, and optical filters, for each wavelength region. For example, MCT (mercury–cadmium–tellurium), PbSe, and Si photo-detectors are used for the mid-infrared (0.06–0.5 eV), near-infrared (0.5–1.5 eV), and visible regions (1.5–3.0 eV), respectively. It is also important in studying the electronic states to measure the absolute value of the reflectivity change. For this purpose, the probe pulse is split into two pulses and these pulses are used to monitor the signals with and without the pump pulse using a pair of photo-detectors. The sample has to be held in a cryostat, which can cool the sample down to a few kelvins, because these samples show a wide variety of phases at low temperatures.

As an example, the red line in Fig. 9.3a shows the transient reflectivity (R) spectrum at 100 fs after photoexcitation with a 1.55 eV pump pulse in an organic strongly correlated material, $(EDO\text{-}TTF)_2PF_6$ (EDO-TTF = 4,5-ethylenedioxytetrathiaful valene), at 180 K [7, 8]. This spectrum is constructed from the sum of the reflectivity change ($\Delta R/R$) spectrum directly obtained by the above method and the static reflectivity spectrum before photoexcitation shown in Fig. 9.3a by the blue line. For comparison, the static reflectivity spectrum at 290 K is shown in Fig. 9.3a by the green line. To perform a K-K transformation, the spectrum has to be measured over as wide

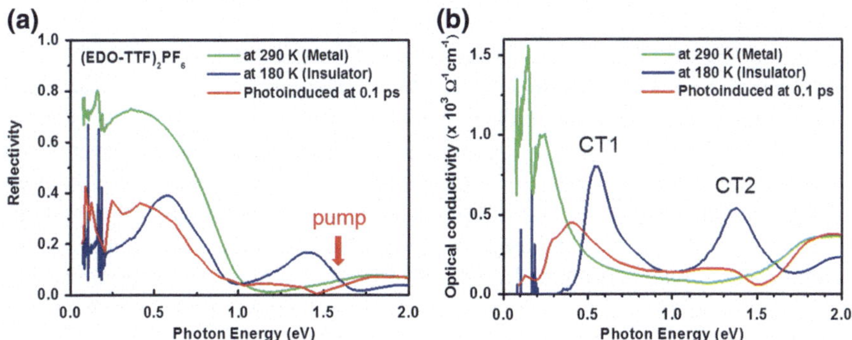

Fig. 9.3 **a** Reflectivity and **b** optical conductivity spectra of $(EDO\text{-}TTF)_2PF_6$. The *blue*, *green*, and *red lines* are spectra at 180, 290 K in thermal equilibrium, and at 0.1 ps after photoexcitation, respectively

a photon energy range as possible because the transformation includes an integration from zero to infinity. The lower photon energy region is especially important for determining the electronic states in these materials, as mentioned above. Thus, we measured the spectrum in Fig. 9.3a down to 0.067 eV (~18 μm) and performed an extrapolation down to 0 eV using an appropriate function. The higher energy side of the spectrum is not so important here because we excite intermolecular states by a near infrared light; thus, we measured the spectrum only up to 2.2 eV and performed an extrapolation using another appropriate function.

Using the K-K transformation, we obtained the optical conductivity spectra from the reflectivity spectra shown in Fig. 9.3b. The optical conductivity (σ) corresponds to the oscillator strength in isolated atoms and molecules and has the following advantages for studying the electronic states in a solid material. The integral of the whole spectrum is unity and the value at zero photon energy is in principle equal to the direct current (DC) conductivity. The red line is the transient optical conductivity spectrum and the blue and green lines are the spectra at 180 and 290 K, respectively. The optical conductivity spectrum at 180 K has a typical character of an insulator phase in a CT complex. The value approaches zero at around zero photon energy, indicating that the sample is an insulator for DC. The broad peaks at 0.55 and 1.38 eV can be attributed to CT transitions between constituent molecules (EDO-TTF for this case). The two peaks indicate that there are two different CT transitions. Various studies [9–11] have shown that EDO-TTF molecules stack one-dimensionally in a crystal as schematically shown in Fig. 9.4, and the order of charges can be represented as (…, 0, +1, +1, 0, …). The transitions at 0.55 and 1.38 eV are assigned to the CTs between EDO-TTF^{0} and EDD-TTF^{+1} and between EDO-TTF^{+1} and EDD-TTF^{+1}. In contrast, the spectrum at 290 K is a typical spectrum of metallic CT complexes. The optical conductivity increases as the photon energy decreases and originates from conduction electrons, and the value extrapolated to zero photon energy agrees with the DC conductivity. In this state, the charge is melting and is expressed as (…, 0.5, 0.5, 0.5, 0.5, …). In short, $(EDO\text{-}TTF)_2PF_6$ is regarded as an insulator at 180 K and a metal at 290 K.

As shown in Fig. 9.3b, the optical conductivity spectrum at 0.1 ps after photoexcitation significantly differs from the spectra at 180 and 290 K and exhibits different features from the phases at 180 and 290 K. The conductivity appears to approach zero as the photon energy decreases, indicating that the photoinduced phase is insulating. The fact that only one CT peak is located at 0.4 eV indicates that charge order is different from the phase at 180 K. For further analysis of the photoinduced phase that has not been observed in thermal equilibrium, theoretical calculations are required. By using a model calculation based on the Hubbard model [12], which is a tight binding model considering the electron–electron interactions, we determined that this charge order is (…, 0, +1, 0, +1, …) as shown in Fig. 9.4, and we assign the peak at 0.4 eV to the CT between EDO-TTF^{0} and EDD-TTF^{+1} [7]. This is the first observation of the photoinduced phase that significantly differs from the phases in a phase diagram in thermal equilibrium. This type of phase is called a hidden phase. Recently another type of dynamical calculation using DFT has been carried out and the further details of this PIPT were investigated [13].

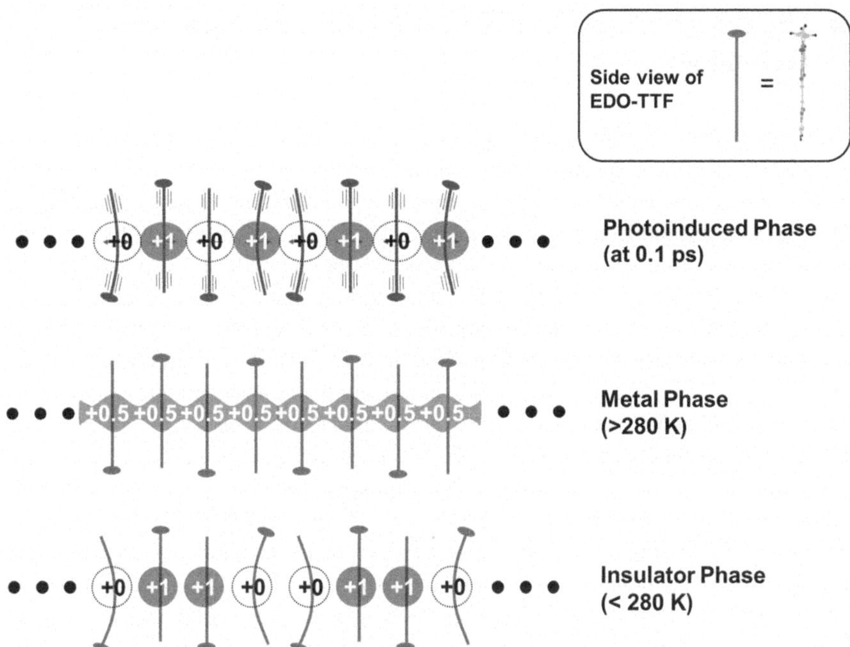

Fig. 9.4 Schematic of the order of EDO-TTF molecules in $(EDO\text{-}TTF)_2PF_6$ in each phase. The number on EDO-TTF represents the charge of EDO-TTF

To date, using this type of technique, several PIPTs of CT complexes have been investigated. TTF-CA (tetrathiafulvalene-p-chloranil) was found to be the first known CT complex showing PIPT in 1990, studied using a nanosecond laser [14]. After 2000, the spectra over a wide photon energy range was measured using a femtosecond Ti:sapphire CPA by several research groups [15–17]. Subsequently, the following PIPTs have been studied. TCNQ salt (TCNQ = tetracyanoquinodimethane) is a one-dimensional 1/2-filled electronic system and shows a typical phase transition based on one-dimensionality [18, 19]. BEDT-TTF salt (BEDT-TTF = bis(ethylenedithio)tetrathiafulvalene) is a well-known CT complex having a two-dimensional electronic system and shows various types of PIPT depending on the crystal structure [20, 21]. $Pd(dmit)_2$ salt (dmit = 1,3-dithiole-2-thione-4,5-dithiolate) forms a complex consisting of strongly bound dimers and exhibits a PIPT consistent with this dimerization [22]. $(EDO\text{-}TTF)_2PF_6$ is a one-dimensional 3/4-filled electronic system and exhibits a strong electron–phonon interaction in addition to an electron–electron interaction [7].

9.4 The Initial Process of PIPT Using Pulse Compression Techniques

As described above, a photoinduced phase emerging up to 100 fs can be identified in the transient electronic spectra over a wide photon energy range and for appropriate model calculations. However, the initial process from the direct excited state (F-C state) to the photoinduced phase is much faster. Thus, a method with higher time resolution is required to observe this process. To obtain an ultrashort pulse for such measurement from a conventional CPA, two methods have been used to date. One is non-collinear optical parametric amplification (NOPA) [23–25] and the other is pulse compression using a gas-filled hollow fiber [26]. Here we introduce the study of the first stage of PIPT in $(EDO\text{-}TTF)_2PF_6$ using the latter technique.

Figure 9.5a shows a schematic of the experimental setup, based on [27]. The source CPA laser is the same as that used in Sect. 9.3. The output of the CPA is focused on a hollow fiber held in a gas cell using a 1000-mm lens. The inner diameter and length of the fiber are 250 μm and 50 cm, respectively. The gas cell is filled with two-atom krypton gas. The output pulse from the fiber has a broad band width (∼0.5 eV) but it is strongly chirped, and thus it was compressed using negative chirp mirrors including higher-order chirp mirrors. The fringe-resolved autocorrelation of the pulse is shown in Fig. 9.5b, and the single pulse duration estimated from this was 12 fs. Using this pulse, we studied the PIPT dynamics of $(EDO\text{-}TTF)_2PF_6$ [26].

Figure 9.6 shows the spectra of the 12 fs pulse and reflectivity of $(EDO\text{-}TTF)_2PF_6$. The peak centered at 1.38 eV is the CT2 band, which was also seen in Fig. 9.3a. We used the whole spectrum of the 12 fs pulse for pump and probe pulses but we selected a part of the spectrum using an optical filter in front of a photodetector. Figure 9.7a shows the temporal profile of the reflectivity change at around 1.65 eV shown by shading in Fig. 9.6. In this profile, there are three characteristic features: a quick increase immediately after photoexcitation, a slow increase following the quick increase, and weak oscillation in the negative delay time.

The time profile except for the negative delay signal was analyzed using the model shown in Fig. 9.7b as follows. The sample was excited to the F-C state by a pulse

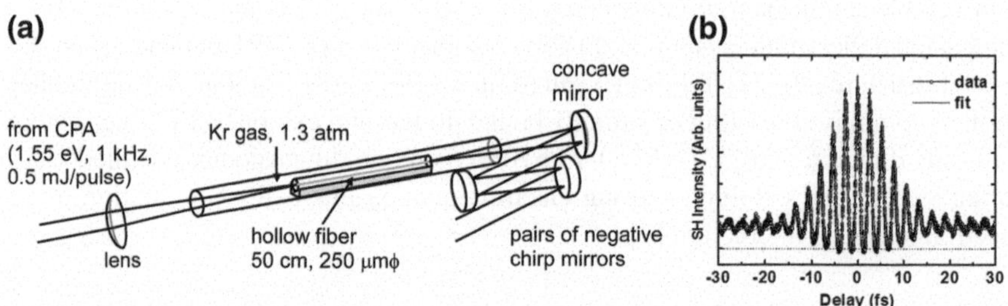

Fig. 9.5 **a** Experimental setup for pulse compression using a gas-filled hollow fiber. **b** Fringe-resolved autocorrelation of the compressed pulse

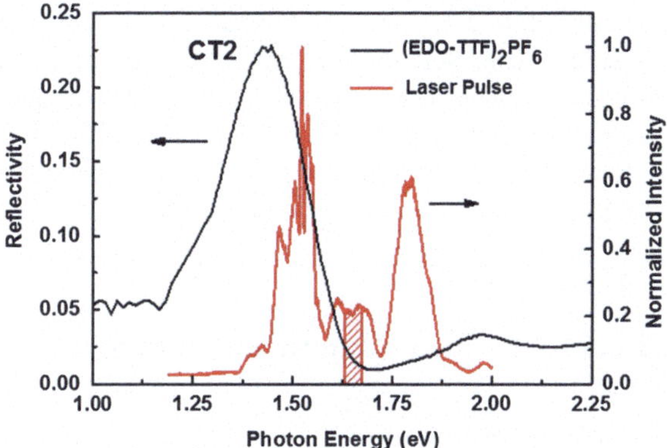

Fig. 9.6 Reflectivity spectrum of $(EDO\text{-}TTF)_2PF_6$ (*black line*) and intensity spectrum of the compressed 12 fs pulse (*red line*)

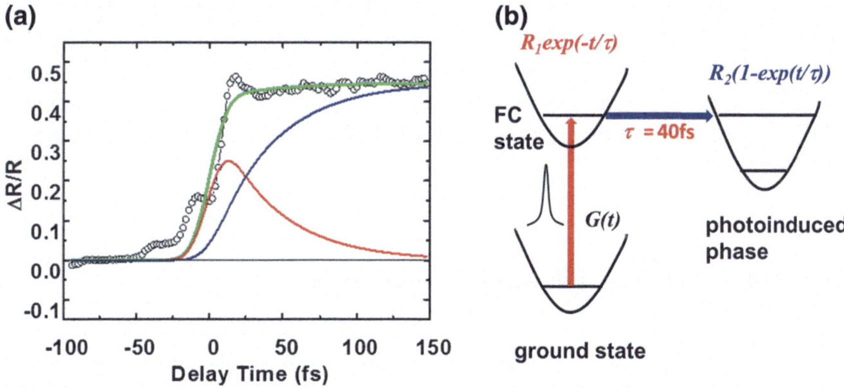

Fig. 9.7 **a** Experimental (*circles*) and simulated (*solid lines*) temporal profiles of reflectivity change after photoexcitation of $(EDO\text{-}TTF)_2PF_6$ at 25 K. **b** Model of the initial process of PIPT using this simulation [26]

expressed by the function G(t). The sample is then changed into the photoinduced phase exponentially with the time constant τ. All the data including those at other probe photon energies were fitted using this model and the time constant $\tau = 40$ fs was obtained. This result indicates that 40 fs is required for the emergence of the photoinduced phase. This delayed emergence of the photoinduced phase is predicted by the theoretical calculation mentioned above and its origin is strong electron–phonon coupling [7].

Other dynamics can be extracted from the negative delay signal. The signal at the negative delay is generated by the third order non-linear optical process [28]. In brief, the detected signal is a part of the pump pulse diffracted by the transient grating made by the pump and probe pulses. This transient grating disappears when the electronic coherence created by the probe pulse decays. Thus, the decay time of the electronic coherence at the F-C state was estimated to be 22 fs from the decay of the negative delay signal. The oscillation accompanying the decay originates from the coherently excited phonons coupled to the F-C state. For a frequency of 38 THz ($\sim$1300 cm^{-1}), the phonon is assigned to the intramolecular vibrations of EDO-TTF molecules. These results suggest that the coherent coupling between electrons and intramolecular vibrations is the precursor state of the PIPT.

9.5 Observation of Ultrafast Structural Change

After photoexcitation of solid materials, the crystal structure does not always vary in the same way as the electronic structure under non-equilibrium conditions. This difference is especially pronounced for organic crystals because of their flexible and soft characters. Thus, time-resolved techniques for probing temporal variations of lattice and molecular structures are important. Currently, systems using a synchrotron radiation light source or X-ray free electron laser are being developed. It is also possible to detect structural changes using table-top systems with ultrashort pulse lasers. We introduce below studies on different dynamics of charge and structure in PIPTs using time-resolved infrared vibrational spectroscopy and time-resolved electron diffraction.

9.5.1 Time-Resolved Vibrational Spectroscopy

Vibrational spectra are used to identify organic materials and as a probe for changes in structure and environment of molecules. They are a powerful tool if the temporal resolution is high enough to resolve crystal structure change after photoexcitation in organic solid materials. In general, this is not an easy technique to apply because of the low absorption coefficient of mid-infrared light (1000–3000 cm^{-1}, 1 cm^{-1} $\sim$ 0.12 meV), where many of the characteristic vibrational peaks of organic molecules are located [29]. Thus, stable and highly sensitive ultrafast measurement systems are required to obtain time-resolved vibrational spectra. This measurement system also requires a high energy resolution, because the peak width of the vibrational transition is generally a few wavenumbers and such peaks are located densely [29]. There are two methods to obtain such high resolution using a 100 fs CPA. One is to generate a narrow band picosecond pulse ($\sim$10 cm^{-1}, $\sim$3 ps) inside the CPA system [30] and the other is to combine a broadband femtosecond pulse ($\sim$150 cm^{-1}, $\sim$120 fs) and a multichannel infrared detector array equipped with a polychromator [31–33]. These two methods both have advantages and disadvantages and thus we have developed both methods to study the dynamics in solid materials.

The experimental setup for the former system is basically the same as that for the transient reflectivity spectrum described in Sect. 9.3, but a narrow-band picosecond pulse was generated using a spatial mask in front of a grating of the pulse stretcher inside a CPA. A tunable narrow-band infrared pulse in the range 1000–4000 cm^{-1} was obtained from the output of the picosecond CPA using OPA and DFG. The pulse energy width and duration were approximately 10 cm^{-1} and 3 ps, respectively. The reflectivity change ($\Delta R/R$) was obtained from the reflected infrared pulse from the sample surface (signal) and a part of the infrared pulse separated by a beam splitter before the sample (reference). If the reference intensity is set to be the signal intensity before photoexcitation of the sample, the exact $\Delta R/R$ can be obtained. This is an advantage of this method using a narrow-band picosecond pulse. Thus, this method is suitable for studying samples having broad peaks over 100 cm^{-1}, which is often observed in solid materials.

Figure 9.8a shows the temporal variation of reflectivity change spectra at 1, 20, and 300 ps in the region of C=C stretching vibrational modes of $(EDO\text{-}TTF)_2PF_6$ after photoexcitation of the CT2 band in the insulating phase at 180 K. Figure 9.8b shows the static reflectivity spectra in the phases at 180 and 290 K measured by FT-IR (Fourier transform infrared spectrometry) [30]. We note that the vibrational spectra were measured for polarization of light perpendicular to the stack direction of EDO-TTF whereas the previous transient electronic spectra were measured for light parallel to the stack direction. At 1 ps after photoexcitation, a reflectivity increase over the whole spectral region and three sharp bleach peaks are observed. Because the wavenumbers of the bleach peaks correspond to the vibrational peaks in the low temperature phase, the bleach peaks indicate a disappearance of the low temperature phase. The broad reflectivity increase indicates a fluctuation of the charge distribution and/or lattice structure in the photoinduced phase. At 20 ps, the broad reflectivity increase decreases, indicating that the charge and lattice fluctuation becomes larger. At 300 ps, there is only a single sharp peak at 1572 cm^{-1}. The wavenumber of this peak is in good agreement with the peak in the spectrum in the phase at 290 K and thus the emergence of the peak seems to indicate emergence of the metallic phase, though interpretation of the reflectivity spectra in solid materials is not simple because of the different penetration depths of the pump and probe pulses. To interpret the spectral change at 300 ps properly, we performed a spectral simulation using the multilayer model as follows [17, 19, 30].

First we assumed that the dielectric constants before and after photoexcitation correspond to the dielectric constants of the phases at 180 and 290 K, respectively, deduced from the reflectivity spectra using the K-K transformation. We also assumed that the density of the photoinduced phase decreases exponentially with penetration depth, d, along the direction of the propagation of the pump light from the surface. The sample is represented as a model composed of many thin layers having the same thickness and different but homogeneous dielectric constants. The dielectric constant of each layer is assumed from the following formula,

$$\varepsilon(x) = \varepsilon^{HT} \gamma \exp\left(-\frac{x}{d}\right) + \varepsilon^{LT} \left\{1 - \gamma \exp\left(-\frac{x}{d}\right)\right\} \tag{9.1}$$

In this formula, the dielectric constant is represented by a linear combination of the dielectric constants of the low (ε^{LT}) and high temperature (ε^{HT}) phases considering

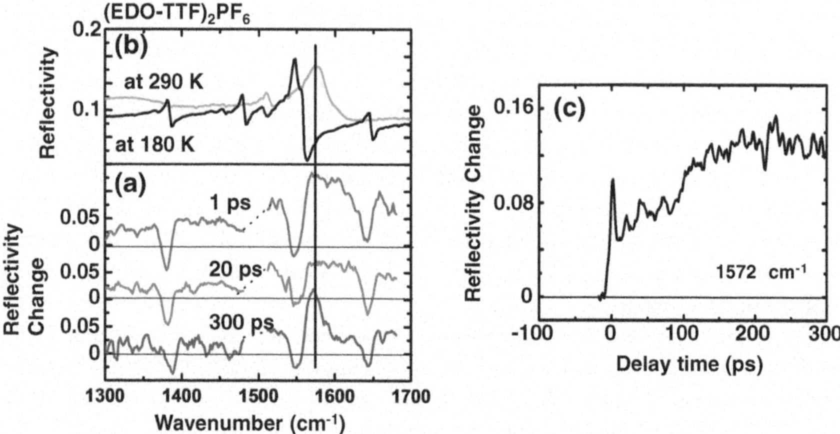

Fig. 9.8 **a** Reflectivity change spectra in the molecular vibration region at 1, 20, and 300 ps after photoexcitation of (EDO-TTF)$_2$PF$_6$. **b** Static reflectivity spectra of (EDO-TTF)$_2$PF$_6$ at 180 and 290 K. **c** Temporal profiles of reflectivity change at 1572 cm^{-1} [30]

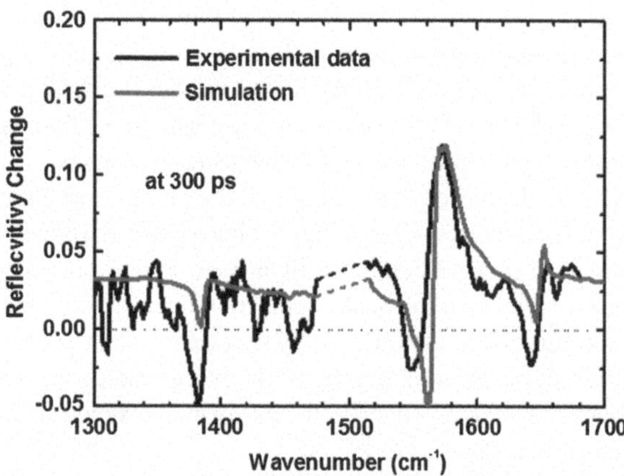

Fig. 9.9 Simulated spectrum using the model described in the text together with the photoinduced reflectivity change spectrum at 300 ps [30]

the distribution of each phase, which is a function of the depth (x) from the sample surface. γ represents the ratio of the photoinduced phase at the sample surface. Under these assumptions, the photoinduced reflectivity spectrum is calculated from the transfer matrix by considering the contributions of each layer. The spectrum simulated by this model is shown by the red line in Fig. 9.9 together with the spectrum at 300 ps. Because the simulated spectrum is in good agreement with the observed spectrum, it is confirmed that the photoinduced phase at least after 300 ps is the same as the high temperature metallic phase.

Figure 9.8c shows the temporal change of the peak at 1572 cm^{-1}. The reflectivity increases immediately after photoexcitation, but this increase does not indicate the emergence of a metallic phase because there is no apparent peak in this time range.

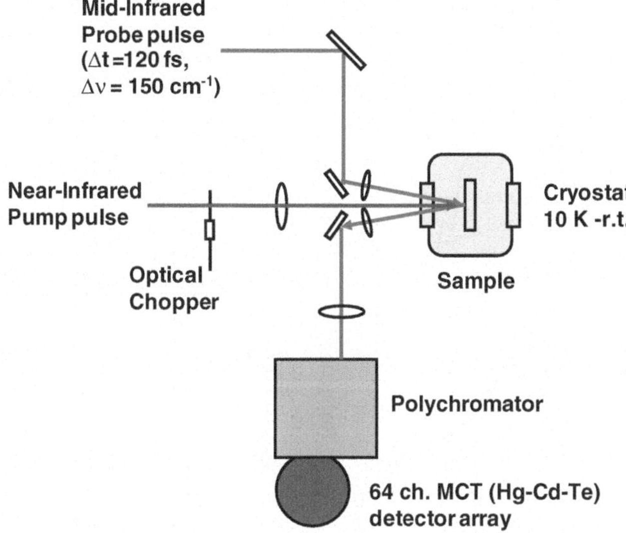

Fig. 9.10 Main part of the experimental setup for time-resolved infrared vibrational spectroscopy using a multi-channel infrared detector array

After a quick decrease, the reflectivity gradually increases over 100 ps. This slow increase does indicate the emergence of a high temperature metallic phase. This result indicates that 100 ps are required for melting of the charge order from the photoinduced state. These dynamics were revealed for the first time by time-resolved vibrational spectroscopy because it has been impossible to follow the process after several picoseconds using time-resolved electronic spectroscopy due to weak and irregular signals [30].

As described so far, time-resolved vibrational spectroscopy using a narrow-band infrared pulse is a powerful tool for observing photoinduced dynamics more than 1 ps in organic solid materials. In particular, this method is effective for samples which scatter laser light, such as solid materials with rough surfaces and powdered materials [34, 35]. However, a disadvantage of this method is that it takes a long time to measure time-resolved spectra because both the energy and delay time have to be scanned. Using a broadband femtosecond infrared pulse, time-resolved vibrational spectra can be acquired more quickly. The energy width of an infrared pulse converted from a 100 fs CPA is approximately 150 cm^{-1}. This width prevents the measurement of vibrational peaks separately. The reflected infrared probe pulse can be dispersed with a grating and detected using a multi-channel infrared detector array, as partially shown in Fig. 9.10. This method is suitable for measuring variations of spectral shape quickly. However, a disadvantage is that it is difficult to obtain exact values of ΔR/R.

An example using this method is the measurement of PIPT in another CT complex, TTF-CA [31]. This sample has two characteristic vibrational peaks: one is $a_g \nu_3$ mode of the TTF molecule and the other is a $b_{1u} \nu_{10}$ mode of CA, as shown in Fig. 9.11a, b,

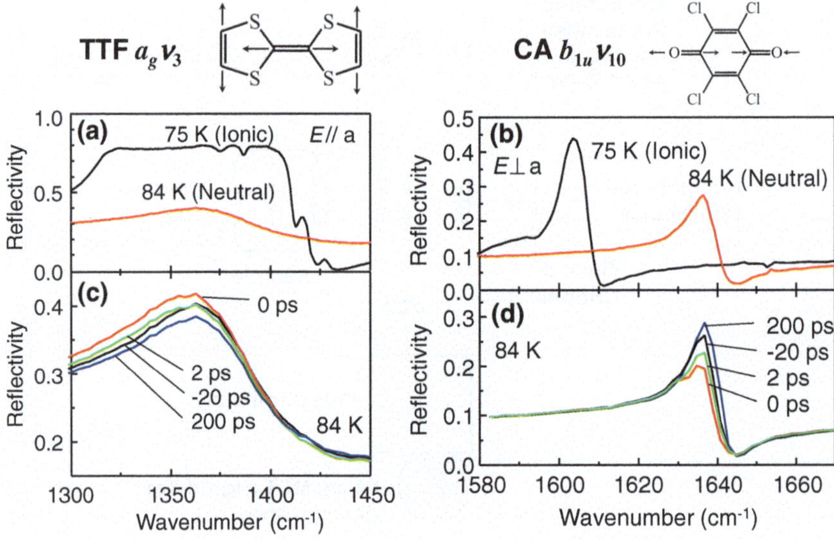

Fig. 9.11 Static reflectivity spectra of TTF-CA at 75 and 84 K in the **a** TTF $a_g \nu_3$ and **b** CA $b_{1u}\nu_{10}$ mode wavenumber region together with the schematic illustration of the vibrations of these modes. **c** and **d** Temporal variations of reflectivity spectra at −20, 0, 2, and 200 ps after photoexcitation in these regions [31]

respectively. It is known that the strength of the $a_g \nu_3$ peak depends on dimerization between TTF and CA molecules, whereas that of the $b_{1u}\nu_{10}$ peak depends on the degree of CT (ρ) between TTF and CA [36]. Thus, if the temporal variations of these peaks are independently obtained, different dynamics of dimerization structure and charge distribution in this material can be revealed separately.

We measured the photoinduced phase transition from the high temperature neutral (N) phase to the low temperature ionic (I) phase. Using a broadband infrared pulse, each peak can be measured without scanning the frequency of the probe pulse. Figure 9.11c, d show the temporal variation of the $a_g \nu_3$ and $b_{1u}\nu_{10}$ peaks, respectively.

The Lorentz model is suitable for extracting physical meaning from these independent bands in reflectivity spectra [6]. This is a phenomenological model that assumes that one harmonic oscillator having a resonant frequency ω_0 gives the dielectric response of a material. The dielectric constant $\varepsilon(\omega)$ is expressed as

$$\varepsilon(\omega) = \varepsilon_\infty + \frac{S\omega_0^2}{\omega_0^2 - \omega^2 + i\gamma\omega}, \tag{9.2}$$

where S, ε_0, and γ are the oscillator strength, dielectric constant above ω_0, and damping constant. The reflectivity is derived from this dielectric constant and all of the obtained reflectivity spectra can be fitted. The insets of Fig. 9.12a, b show an example fitting curve (solid line) and data points (open circles) for the $a_g \nu_3$ and

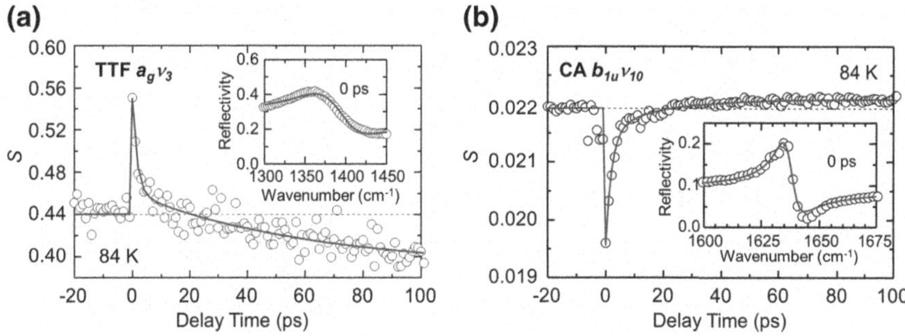

Fig. 9.12 Temporal change in oscillator strength of the **a** TTF $a_g\nu_3$ and **b** CA $b_{1u}\nu_{10}$ mode, obtained using the Lorentz mode. The *insets* are the reflectivity spectrum (*open circles*) and fitting curve (*solid line*) at 0 ps [31]

$b_{1u}\nu_{10}$ peaks, respectively. The curves are in good agreement with the reflectivity spectra. Figure 9.12a, b show the temporal evolutions of S derived at each delay time for the $a_g\nu_3$ and $b_{1u}\nu_{10}$ peaks, respectively. To obtain the time constants of each variation of S, we assumed a double exponential function and obtained time constants of ~2 and ~64 ps for the $a_g\nu_3$ peak, which is the index for dimerization between TTF and CA molecules, and ~2 and ~18 ps for the $b_{1u}\nu_{10}$ peak, which is an index for CT between these molecules.

From these results we determined the different dynamics between structure and charge, as shown in Fig. 9.13. The photoexcitation of CT between TTF and CA in the N phase creates the I phase within 1 ps, which is also observed by the transient reflectivity spectra in a wide photon energy range [17]. The photoinduced I phase partially reverts to the neutral phase in 2 ps, but the other part is changed into the dimeric neutral phase, where the charge is the same as in the neutral phase while the dimerization structure is still the same as that in the I phase. This phase is converted into the N phase over 64 ps accompanied by relaxation of the dimer structure. This result indicates that the structural relaxation is also delayed from the charge relaxation.

9.5.2 Time-Resolved Electron Diffraction

Using time-resolved vibrational spectroscopy, different charge and structure dynamics in solid materials can be investigated. However, the information on structure extracted from the vibrational spectra is indirect. A general method to observe structure in real space is to use X-ray or electron diffraction. To date, considerable effort has been made to obtain time-resolved diffraction patterns, mainly using X-ray pulses of synchrotron radiation. However, the weak X-ray scattering power of organic crystals prevents us from obtaining clear patterns. The scattering power of electron beams is much higher than that of X-rays and thus they are more suitable for time-resolved

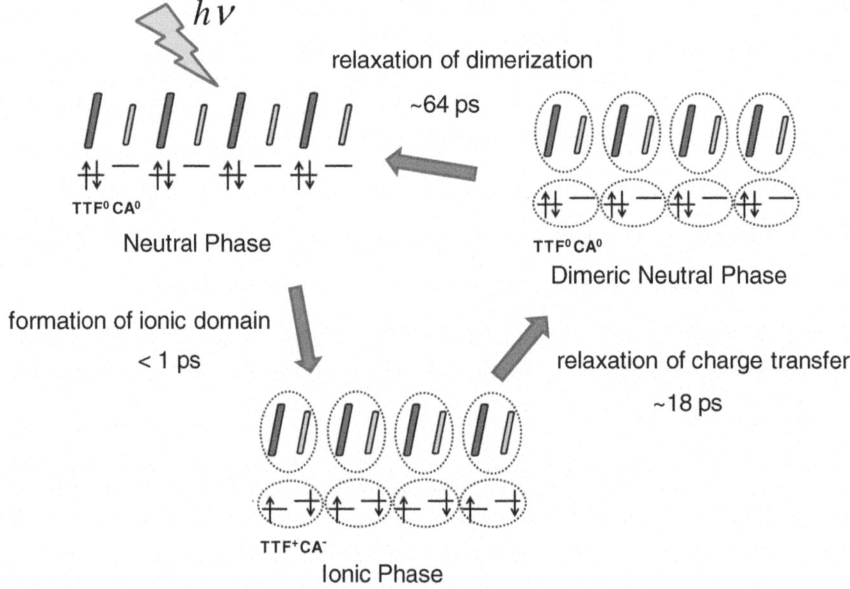

Fig. 9.13 Schematic summary of the photoinduced dynamics in TTF-CA

studies of organic materials. An ultrashort electron pulse for this purpose can be generated from a metal foil excited by an ultrashort pulse whose photon energy is just above the work function of the metal. If the metal is excited by a 100 fs pulse, a 100 fs electron pulse is obtained immediately after the excitation. However, such an electron pulse becomes spatially extended after a while due to Coulomb repulsion between electrons, and the temporal width also increases. To compress this electron pulse, the longitudinal momentum-position distribution of electrons has to be reversed using electric or magnetic fields [37], and several time-resolved electron diffraction systems are being developed to overcome this problem [38–42].

Here we introduce the first observation of structural change in an organic solid crystal using time-resolved electron diffraction [43]. An ultrashort electron pulse is obtained from a 20 nm thickness of gold foil excited by the third harmonic (4.7 eV) of an output of a CPA (50 fs, 800 nm, 1 kHz). The electron pulse is compressed by RF (radio frequency cavity), which makes the delayed electrons faster and the advanced ones slower using a synchronous high frequency electrical field. The sample was sliced in 100 nm so that the electron bunch passes through the sample, and is attached on a copper mesh to avoid charging of the sample. The diffraction pattern of electrons through the sample was detected with CCD camera. The sample was excited by a part of the output of the CPA and time-resolved diffraction patterns were obtained. The temporal instrument response function of the system was 430 fs [39].

Figure 9.14a, b show the difference diffraction patterns at 100 ps after photoexcitation and before photoexcitation of $(EDO\text{-}TTF)_2PF_6$ at 230 K and the temporal

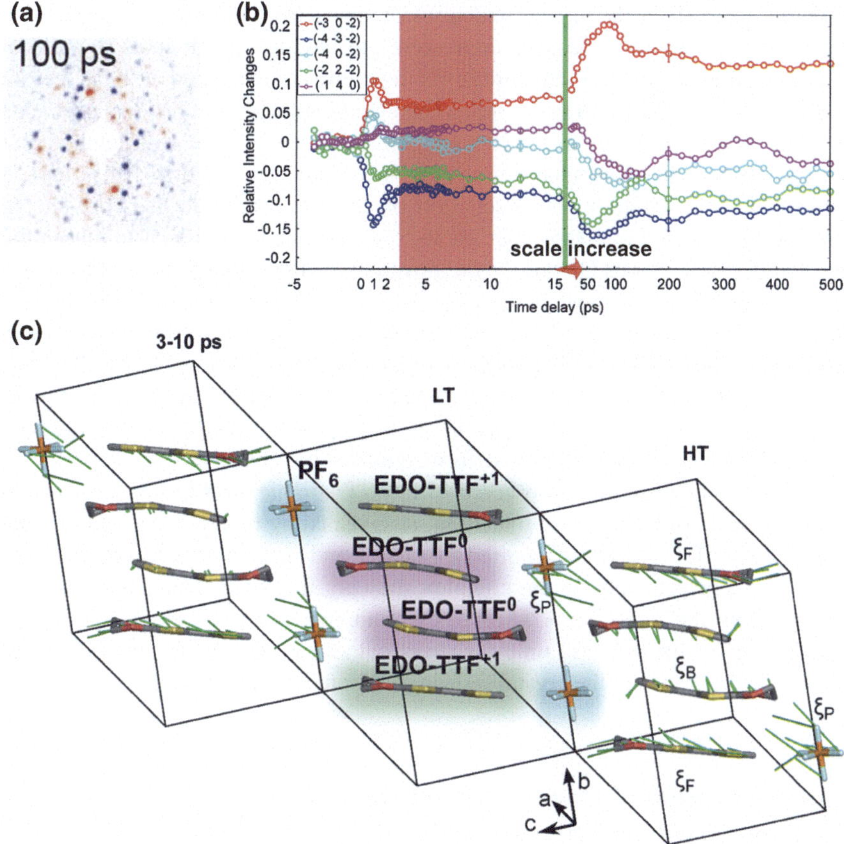

Fig. 9.14 **a** Difference diffraction pattern at 100 ps after photoexcitation and before photoexcitation of $(\text{EDO-TTF})_2 PF_2$ at 230 K. **b** Temporal variation of relative intensity change of the selected diffraction spots. **c** Crystal structure of $(\text{EDO-TTF})_2 PF_6$ before photoexcitation (*middle*), displacements of atoms represented by *green lines* at the phase transition from 230 (LT) to 295 K (HT) (*right*), and those at the photoinduced phase transition at 3–10 ps after photoexcitation (*left*) [43]

variation of the relative intensity change of the selected diffraction spots, respectively. The variation of the relative intensity for all the selected diffraction spots increases immediately after photoexcitation and then decreases within a picosecond. Subsequently, the variation increases slowly over ~100 ps. This temporal variation is in good agreement with that obtained by time-resolved vibrational spectroscopy, shown in Fig. 9.8c, which proves that both the variations of diffraction spots and vibrational peaks reflect the same dynamics of structural change.

Figure 9.14c show the crystal structures obtained from the diffraction patterns, under some assumptions because there were too few diffraction spots to perform a full crystal analysis. The structure in the low temperature phase at 230 K before photoexcitation is shown in the middle. The green bars represent the displacements from the atomic positions in the low temperature phase for the high temperature phase at 290 K on the right hand side and the photoinduced phase at 3–10 ps (marked in red in Fig. 9.14b) after photoexcitation on the left hand side. It is not shown here but the structure at 100 ps after photoexcitation becomes similar to that in the high temperature phase. Obviously the positions of the EDO-TTF^{+1} and PF_6^- in the photoinduced phase move in the same direction to those in the high temperature phase, while that of the EDO-TTF^0 does not move. A remarkable difference between EDO-TTF^{+1} and EDO-TTF^0 is their molecular structures, that is, EDO-TTF^{+1} is flat and EDO-TTF^0 is bent. Thus, it is likely that EDO-TTF^0 requires 100 ps to transform into the same structure as the high temperature phase because the bent EDO-TTF^0 has to be more greatly deformed during the process.

A similar phenomenon is observed in the CT complex Pd(dmit)$_2$ salts, which have a different order of phase transition, as revealed by time-resolved vibrational spectroscopy [33]. The times required for charge and structural variations are <0.1 and ~70 ps, respectively, in the complex showing first-order phase transition. In contrast, both the times are <0.1 ps in the complex showing second-order phase transition. This difference in time for structural change is considered to originate from steric hindrance in the crystal showing first-order phase transition.

9.6 Observation of Domain Structure Change Using Time-Resolved Microscopy

Using the methods described so far, most of the photoinduced dynamics in organic crystals can be revealed. However, one more important piece of information remains, that is, temporal variation of the domain structure. In general, a phase transition takes place inhomogeneously and growth of the domain structure can be observed [44]. Thus, it is possible that variation of the domain structure plays an important role during photoinduced phase transition [45]. However, no direct observations of temporal variation of domain structure have been made to date. The size of such domains is expected to be sub-micrometer and thus such a spatial resolution is required for the measurement system in addition to a femtosecond temporal resolution. One of the best methods to realize this is TR-PEEM (time-resolved photoemission electron microscopy). PEEM is an electron microscope which acquires an image of electrons emitted by light irradiation using electron lenses. If the photoemission electrons are generated by an ultrashort pulse, an ultrashort snapshot of the sample can be obtained

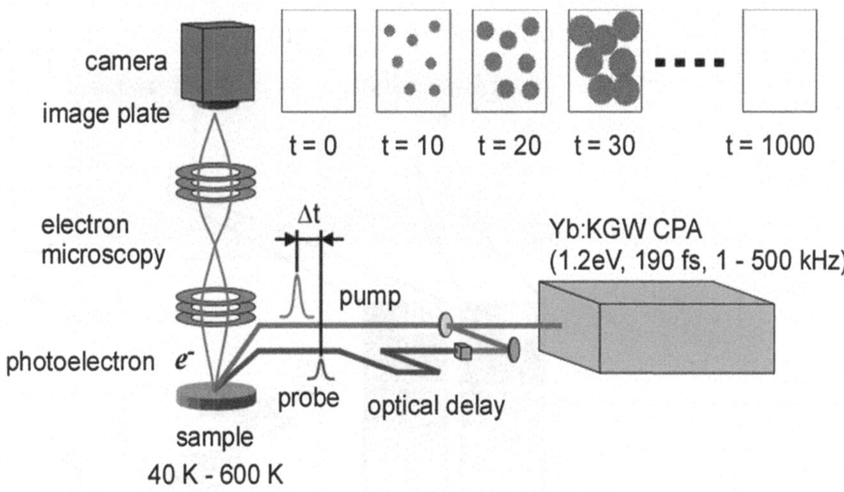

Fig. 9.15 Experimental setup for obtaining photoinduced images with sub-nanometer spatial and sub-picosecond temporal resolution using a PEEM and a Yb:KGW laser

using the pump-probe method. TR-PEEM has already been used for some nanostructures of metals or semiconductors [46–48]. However, such systems cannot be applied to the study of PIPT for the following reasons.

Conventional TR-PEEM works only at room temperature, while phase transitions take place when the temperature varies. Electrons generated by an ultrashort pulse are packed in a small space, allowing for a lower pulse energy and higher repetition rate. However, PIPT has a threshold for the excitation pulse and the long lifetime of photoinduced phase prevents a high repetition rate. To address these problems, we have developed a novel TR-PEEM system shown in Fig. 9.15 [49, 50]. The light source is a CPA using a Yb:KGW crystal. The photon energy, pulse duration, and power of the CPA are 1.2 eV, 190 fs, and 6 W, respectively. The repetition rate is variable from 1 to 500 kHz thanks to the heat characteristic of the KGW crystal. This feature allows us to find the best repetition rate to observe each sample. Moreover, using a highly stable sample holder, the sample temperature can be varied from 40 to 600 K without thermal drift.

Using this system, we have first observed the movement of photocarriers in a simple semiconductor, GaAs [49]. Figure 9.16a, b show the images of a GaAs wafer with copper electrodes at 20 and 40 ps after photoexcitation with a 2.4 eV pulse, respectively. The white strip in the center of each image is the spatial distribution of photocarriers created by the photoexcitation. We could obtain clear images at a 50 kHz repetition rate and not at lower or higher repetition rates. We moved the photocarriers by applying an electric field of 2000 V/cm between the electrodes. To show the movements easily, the areas of the red and blue rectangles are expanded and shown in Fig. 9.16c, d, respectively. The intensity distributions of these rectangle regions are plotted in Fig. 9.16e. The center position of the white stripe at 40 ps is

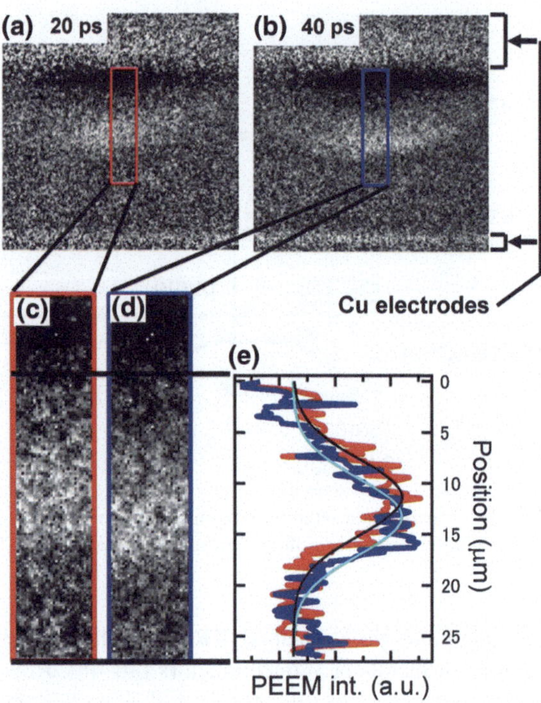

Fig. 9.16 **a** and **b** TR-PEEM images of a GaAs wafer with copper electrodes at 20 and 40 ps after photoexcitation with a 2.4 eV pulse, respectively. **c** and **d** Enlarged views of the *rectangular regions* in (**a**) and (**b**). **e** Intensity profiles of the *red* and *blue rectangular regions* [49]

located ~2 μm below that at 20 ps, corresponding to a velocity of roughly 5×10^6 cm/s. We estimated the mobility of GaAs to be 3300 $cm^2\ V^{-1}\ s^{-1}$ by measuring velocities under various electronic fields. This result indicates that a variable repetition rate works well for obtaining images of carriers having a long lifetime.

Even using a variable repetition rate, it is hard to obtain images of the insulator phase. After photoemission takes place, electrons must be supplied to the sample, otherwise the residual charge in the sample distorts the PEEM image. A sample with low electrical conductivity is not readily re-supplied. We addressed this problem by depositing copper in a stripe pattern on a sample through a micrometer-scale mask. Figure 9.17 shows the image of $(EDO\text{-}TTF)_2PF_6$ at 265 K. Although the material is in the insulator phase at this temperature, we obtained clear images. It is noted that the dark stripes are the edges of the deposited copper. We found from this image that the crystal is broken into small flakes due to a large volume change accompanying the first-order phase transition. By using this method, we are now collecting images under photoinduced phase transitions.

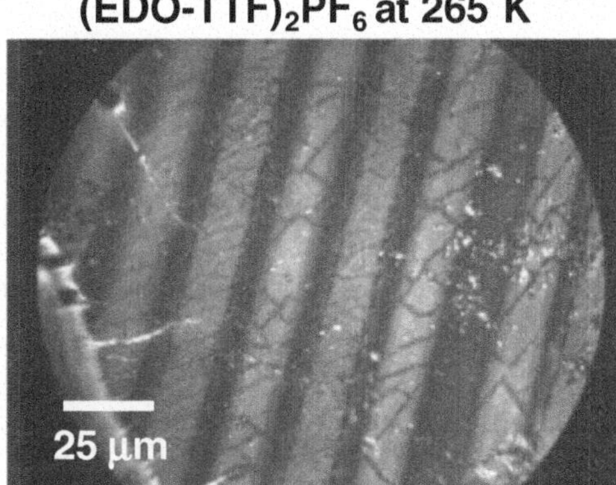

Fig. 9.17 PEEM image of insulating $(EDO\text{-}TTF)_2PF_6$ at 265 K

9.7 Concluding Remarks

We have reviewed a wide variety of photoinduced dynamics in strongly correlated organic crystals and various ultrafast techniques for studying these dynamics. To understand the transient electronic states, techniques with a wide temporal range from 10 fs to nanoseconds and with a wide photon energy range from far infrared to visible were required. In addition, time-resolved vibrational spectroscopy, electron-diffraction, and microscopy were useful for understanding ultrafast microscopic and macroscopic structure changes. Using these techniques, we found that various photoinduced states emerge step by step after photoexcitation, originating from different time scales of electronic and structural changes. If we can understand and control these phenomena, it would be possible to control physical properties such as photoconductivity and magnetism on the femtosecond to picosecond time scales. Moreover, their cooperative phenomena based on strong electron–electron and electron–phonon interactions would be applicable for high efficient photoenergy conversion. However, these complex non-equilibrium phenomena in solid materials are still far from being fully understood and further developments of ultrafast techniques for addressing this are required.

Acknowledgments The author wishes to thank Prof. Shin-ya Koshihara, Prof. Yoichi Okimoto, Dr. Tadahiko Ishikawa, Dr. Keiki Fukumoto (Tokyo Institute of Technology), Hideki Yamochi (Kyoto University), and Prof. R.J. Dwayne Miller (University of Toronto) for their valuable support and contributions to this work.

References

1. E. Dagotto, Science **309**, 257 (2005)
2. K. Nasu, *Photo Induced Phase Transition* (World Scientific, Singapore, 2003)
3. M. Gonokami, S. Koshihara (eds.), Special issue on photo-induced phase transitions and their, dynamics. J. Phys. Soc. Jpn. **75**, 011001 (2006)
4. S. Kagoshima, K. Kanoda, T. Mori (eds.), Special topics on organic conductors. J. Phys. Soc. Jpn. **75**, 051001 (2006)
5. S. Uji, T. Mori, T. Takahashi (eds.), Topical review on focus on organic conductors. Sci. Technol. Adv. Mater. **10**, 020301 (2009)
6. For example, M. Fox, *Optical Properties of Solids*, (Oxford University Press, Oxford, 2001)
7. K. Onda, S. Ogihara, K. Yonemitsu, N. Maeshima, T. Ishikawa, Y. Okimoto, X.F. Shao, Y. Nakano, H. Yamochi, G. Saito, S. Koshihara, Phys. Rev. Lett. **101**, 067403 (2008)
8. K. Onda, S. Ogihara, T. Ishikawa, Y. Okimoto, X.F. Shao, H. Yamochi, G. Saito, S. Koshihara, J. Phys. Cond. Mat. **20**, 224018 (2008)
9. A. Ota, H. Yamochi, G. Saito, J. Mater. Chem. **12**, 2600 (2002)
10. O. Drozdova, K. Yakushi, A. Ota, H. Yamochi, G. Saito, Synth. Met. **133–134**, 277 (2003)
11. O. Drozdova, K. Yakushi, K. Yamamoto, A. Ota, H. Yamochi, G. Saito, H. Tashiro, D.B. Tanner, Phys. Rev. B **70**, 075107 (2004)
12. K. Yonemitsu, N. Maeshima, Phys. Rev. B **76**, 075105 (2007)
13. K. Iwano, Y. Shimoi, Phys. Rev. Lett. **110**, 116401 (2013)
14. S. Koshihara, Y. Tokura, T. Mitani, G. Saito, T. Koda, Phys. Rev. B **44**, 6853 (1990)
15. S. Iwai, S. Tanaka, K. Fujinuma, H. Kishida, H. Okamoto, Y. Tokura, Phys. Rev. Lett. **88**, 057402 (2002)
16. K. Tanimura, Phys. Rev. B **70**, 144112 (2004)
17. H. Okamato, Y. Ishige, S. Tanaka, H. Kishida, S. Iwai, Y. Tokura, Phys. Rev. B **70**, 165202 (2004)
18. H. Okamoto, K. Ikegami, T. Wakabayashi, Y. Ishige, J. Togo, H. Kishida, H. Matsuzaki, Phys. Rev. Lett. **96**, 037405 (2006)
19. H. Okamoto, H. Matsuzaki, T. Wakabayashi, Y. Takahashi, T. Hasegawa, Phys. Rev. Lett. **98**, 037401 (2007)
20. S. Iwai, K. Yamamoto, A. Kashiwazaki, F. Hiramatsu, H. Nakaya, Y. Kawakami, K. Yakushi, H. Okamoto, H. Mori, Y. Nishio, Phys. Rev. Lett. **98**, 097402 (2007)
21. Y. Kawakami, S. Iwai, T. Fukatsu, M. Miura, N. Yoneyama, T. Sasaki, N. Kobayashi, Phys. Rev. Lett. **103**, 066403 (2009)
22. T. Ishikawa, N. Fukazawa, Y. Matsubara, R. Nakajima, K. Onda, Y. Okimoto, S. Koshihara, M. Lorenc, E. Collet, M. Tamura, R. Kato, Phys. Rev. B **80**, 115108 (2009)
23. H. Uemura, H. Okamoto, Phys. Rev. Lett. **105**, 258302 (2010)
24. Y. Kawakami, T. Fukatsu, Y. Sakurai, H. Unno, H. Itoh, S. Iwai, T. Sasaki, K. Yamamoto, K. Yakushi, K. Yonemitsu, Phys. Rev. Lett. **105**, 246402 (2010)
25. S. Wall, D. Brida, S.R. Clark, H.P. Ehrke, D. Jaksch, A. Ardavan, S. Bonora, H. Uemura, Y. Takahashi, T. Hasegawa, H. Okamoto, G. Cerullo, A. Cavalleri, Nat. Phys. **7**, 114 (2011)
26. Y. Matsubara, S. Ogihara, J. Itatani, N. Maeshima, K. Yonemitsu, T. Ishikawa, Y. Okimoto, S. Koshihara, T. Hiramatsu, Y. Nakano, H. Yamochi, G. Saito, K. Onda, Phys. Rev. B **89**, 161102(R) (2014)
27. M. Nisoli, S. De Silvestri, O. Svelto, Appl. Phys. Lett. **68**, 2793 (1996)
28. T. Kobayashi, J. Du, W. Feng, K. Yoshino, Phys. Rev. Lett. **101**, 037402 (2008)
29. M.D. Fayer (ed.), *Ultrafast Infrared Vibrational Spectroscopy* (CRC Press, Boca Raton, 2013)
30. N. Fukazawa, M. Shimizu, T. Ishikawa, Y. Okimoto, S. Koshihara, T. Hiramatsu, Y. Nakano, H. Yamochi, G. Saito, K. Onda, J. Phys. Chem. C **116**, 5892 (2012)
31. Y. Matsubara, Y. Okimoto, T. Yoshida, T. Ishikawa, S. Koshihara, K. Onda, J. Phys. Soc. Jpn. **80**, 124711 (2011)
32. Y. Matsubara, T. Yoshida, T. Ishikawa, Y. Okimoto, S. Koshihara, K. Onda, Acta Physica Polonica A **121**, 340 (2012)

33. N. Fukazawa, T. Tanaka, T. Ishikawa, Y. Okimoto, S. Koshihara, T. Yamamoto, M. Tamura, R. Kato, K. Onda, J. Phys. Chem. C **117**, 13187 (2013)
34. K. Onda, K. Tanabe, H. Noguchi, A. Wada, T. Shido, A. Yamaguchi, Y. Iwasawa, J. Phys. Chem. B **105**, 11456 (2001)
35. K. Onda, K. Tanabe, H. Noguchi, A. Wada, K. Domen, C. Hirose, J. Phys. Chem. B **107**, 11391 (2003)
36. S. Horiuchi, Y. Okimoto, R. Kumai, Y. Tokura, J. Phys. Soc. Jpn. **69**, 1302 (2000)
37. M. Hada, K. Pichugin, G. Sciaini, Eur. Phys. J. Special Topics **222**, 1093 (2013)
38. B.J. Siwick, A.A. Green, C.T. Hebeisen, R.J.D. Miller, Opt. Lett. **30**, 1057 (2005)
39. M. Gao, H. Jean-Ruel, R.R. Cooney, J. Stampe, M. de Jong, M. Harb, G. Sciaini, G. Moriena, R.J. Dwayne, Miller. Opt. Express **20**, 12048 (2012)
40. T. van Oudheusden, P.L.E.M. Pasmans, S.B. van der Geer, M.J. de Loos, M.J. van der Wiel, O.J. Luiten, Phys. Rev. Lett. **105**, 264801 (2010)
41. A. Gahlmann, S.T. Park, A.H. Zewail, Phys. Chem. Chem. Phys. **10**, 2894 (2008)
42. M. Aidelsburger, F.O. Kirchner, F. Krausz, P. Baum, Proc. Natl. Acad. Sci. **107**, 19714 (2010)
43. M. Gao, C. Lu, H. Jean-Ruel, L.C. Liu, A. Marx, K. Onda, S. Koshihara, Y. Nakano, X. Shao, T. Hiramatsu, G. Saito, H. Yamochi, R.R. Cooney, G. Moriena, G. Sciaini, R.J.D. Miller, Nature **496**, 343 (2013)
44. H. Nishimori, G. Ortiz, *Elements of Phase Transitions and Critical Phenomena* (Oxford University Press, Oxford, 2011)
45. S. Iwai, J. Lumi. **131**, 409 (2011)
46. O. Schmidt, M. Bauer, C. Wiemann, R. Porath, M. Scharte, O. Andreyev, G. Sch€onhence, M. Aeschlimann. Appl. Phys. B **74**, 223 (2002)
47. A. Kubo, K. Onda, H. Petek, Z. Sun, Y.S. Jung, H.K. Kim, Nano Lett. **5**, 1123 (2005)
48. M. Aeschlimann, M. Bauer, D. Bayer, T. Brixner, S. Cunovic, F. Dimler, A. Fischer, W. Pfeiffer, M. Rohmer, C. Schneider, F. Steeb, C. Strüber, D. Voronine, Proc. Natl. Acad. Sci. **107**, 5329 (2010)
49. K. Fukumoto, Y. Yamada, K. Onda, S. Koshihara, Appl. Phys. Lett. **104**, 053117 (2014)
50. K. Fukumoto, K. Onda, Y. Yamada, T. Matsuki, T. Mukuta, S. Tanaka, S. Koshihara, Rev. Sci. Inst. **85**, 083705 (2014)

Index

K. Yamanouchi et al. (eds.), *Progress in Ultrafast Intense Laser Science XII*,
Springer Series in Chemical Physics 112, DOI 10.1007/978-3-319-23657-5

The manufacturer's authorised representative in the EU is Springer Nature Customer Service Centre GmbH, Europaplatz 3, 69115 Heidelberg, Germany. If you have any concerns regarding our products, please contact ProductSafety@springernature.com

Printed and bound by CPI Group (UK) Ltd, Croydon, CR0 4YY

15/07/2026

02167635-0001